Diagnostic Techniques
in Genetics

Diagnostic Techniques in Genetics

Edited by
Jean-Louis Serre
Université de Versailles

Translated by
Isabelle and Simon Heath

John Wiley & Sons, Ltd

First published in French as *Analyse de Génomes, Les Diagnostics Génétiques* © 2002 Dunod, Paris

Translated into English by Isabelle & Simon Heath.

This work has been published with the help of the French Ministère de la Culture-Centre national du livre

English language translation Copyright © 2006 John Wiley & Sons Ltd,
The Atrium, Southern Gate, Chichester,
West Sussex PO19 8SQ, England

Telephone   (+44) 1243 779777

Email (for orders and customer service enquiries): cs-books@wiley.co.uk
Visit our Home Page on www.wileyeurope.com or www.wiley.com

Other Wiley Editorial Offices

John Wiley & Sons Inc., 111 River Street, Hoboken, NJ 07030, USA

Jossey-Bass, 989 Market Street, San Francisco, CA 94103-1741, USA

Wiley-VCH Verlag GmbH, Boschstr. 12, D-69469 Weinheim, Germany

John Wiley & Sons Australia Ltd, 33 Park Road, Milton, Queensland 4064, Australia

John Wiley & Sons (Asia) Pte Ltd, 2 Clementi Loop #02-01, Jin Xing Distripark, Singapore 129809

John Wiley & Sons Canada Ltd, 6045 Freemont Blvd, Mississauga, Ontario, Canada L5R 413

Wiley also publishes its books in a variety of electronic formats. Some content that appears in print may not be available in electronic books.

Library of Congress Cataloging-in-Publication Data

Diagnostics génétiques. English.
Diagnostic techniques in genetics / edited by Jean-Louis Serre ;
translated by Isabelle and Simon Heath.
p.   cm.
Includes bibliographical references and index.
ISBN-13: 978-0-470-87024-2 (cloth : alk. paper)
ISBN-10: 0-470-87024-9 (cloth : alk. paper)
ISBN-13: 978-0-470-87025-9 (pbk. : alk. paper)
ISBN-10: 0-470-87025-7 (pbk. : alk. paper)
1. Molecular genetics.   2. Molecular diagnosis.   3. Genetic screening.
I. Serre, Jean-Louis.   II. Title.
[DNLM: 1. Genetic Techniques.   2. Molecular Diagnostic Techniques–methods.
3. Genetic Screening–methods.   QZ 52 D536 2006a]
QH442.D5313 2006
616'.042–dc22   2006010655

British Library Cataloguing in Publication Data

A catalogue record for this book is available from the British Library

ISBN-13 978-0-470-87024-2 (HB)   ISBN-13 978-0-470-87025-9 (PB)
ISBN-10 0-470-87024-9 (HB)   ISBN-10 0-470-87025-7 (PB)

Typeset in 10.5/13pt Minion by TechBooks, New Delhi, India
Printed and bound in Great Britain by Antony Rowe Ltd., Chippenham, Wilts
This book is printed on acid-free paper responsibly manufactured from sustainable forestry
in which at least two trees are planted for each one used for paper production.

Contents

Preface

During the second half of the 20th century, in less than 50 years, there have been two revolutions in biology – one in protein biochemistry followed by one in molecular genetics – with major repercussions both for fundamental research and medical applications.

The development of protein biochemistry (at the end of the 1950s) has led to the purification of many proteins followed by the study – using crystallography – of their three-dimensional structure and, in parallel with this, their sequencing (identification of the linking of amino acids).

The study of variants for a peptide chain then allowed – by identification of amino acid substitutions, notably those caused by missense mutations in the coding sequence of a gene – the exhaustive definition of the causal relationships between each mutation, its molecular effect and its cellular, tissular and clinical effects in many pathologies. Of these, the most intensely studied relationship has been for the protein haemoglobin in the haemoglobinopathies.

Following this first revolution there were many medical applications, but these were unfortunately limited by the fact that the study of proteins can only be the study of gene products, which requires that these products are present and accessible for analysis.

The best example of this is the prenatal diagnosis of β-thalassaemia, a lethal haemoglobinopathy resulting from the absence of haemoglobin β-chains in children having loss of function mutations in both of their copies of the β-gene. The diagnosis was already challenging because it required taking a fetal blood sample *in utero* from a vein in the cord without contaminating it with maternal blood and without provoking a miscarriage and, above all, any success was tempered by the fact that because transcription of the β-gene only starts replacing that of the γ-genes in the fourth month of pregnancy, biochemical tests of the presence or absence of adult haemoglobin ($\alpha2\beta2$) in the essentially entirely fetal haemoglobin ($\alpha2\lambda2$) are pushed back to the fifth or sixth month, leading necessarily to late pregnancy terminations with obvious psychological, clinical and ethical consequences.

The consequences of the second revolution – in molecular genetics – have been more important in the domain of fundamental research than in its applications, simply because it gives direct access to the genes which direct biological phenomena whereas before it was only possible, at best, to access their products or just the phenotypic perturbations of the phenomena resulting from mutations.

Molecular genetics has allowed the cloning of genes, their sequencing and, above all, the ability to isolate them *in vitro*, in vectors, to modify them in a targeted manner, and to reintroduce them into cells or organisms to study, in a controlled manner, the way in which they control one or other aspect of a phenomenon.

Genetics, in becoming molecular, has very rapidly become an integral part of most biological fields, not only due to its power to dissect a biological phenomenon but also because of its increasing ability to unify disciplines which had previously appeared different or unrelated.

In all disciplines such as physiology, embryology and oncology, development and aging, cell biology, population biology, ecology and evolution, the phenomena in question are associated with the expression of specific genes, which can now be localized and identified, cloned and modified at leisure by directed mutagenesis, and so the effects can be studied in a chosen context, *in vitro, ex* or *in vivo*. Molecular genetics is therefore a remarkable tool because potentially it allows the opening of all the 'black boxes' that in these disciplines represent pertinent, but essentially descriptive, approaches without taking account of the global view of the phenomena or the studied structures.

For human genetic pathologies, the molecular and cellular description of the physiopathology of a disease is the condition sine qua non for therapeutic research and, in future, for the possible development of pre- or post-natal genetic diagnosis or even genetic screening of heterozygotic carriers for recessive diseases.

Therefore, the question of the diagnosis of β-thalassaemia can nowadays be resolved without difficulty since direct study of the β-genes from fetal DNA shows whether the two copies are mutated or not in the first weeks of a pregnancy, well before protein biochemistry could indicate the presence or absence of haemoglobin A.

However, genetic diagnoses – mainly pre-natal for severe pathologies – are only one of the principal clinical applications of molecular biology. The molecular biology of DNA has also revolutionized many approaches such as the diagnosis of infectious diseases where the pathogenic agent, bacterium or virus, will be identified by the presence of sequences specific to its genome, or the traceability studies that can now be performed by searching for transgenic elements identifying genetically modified organisms (GMOs).

Even the screening for anomalies in chromosome number (monosomies, trisomies) that is normally performed by karyotyping, which is a classic and highly reliable technique, can sometimes be performed using quantitative techniques for estimating molecular dosage much faster (24 to 48 hours rather than 10 to 15 days) and much more cheaply, with nevertheless some limitations which will be described in this book.

In addition, some karyotypic anomalies, not accessible to conventional cytogenetics (microdeletions, uniparental disomies) can be detected or identified by techniques such as fluorescent *in situ* hybridization (FISH) combining karyotyping and molecular biology.

Oncology itself relies heavily on molecular biology which has allowed, at the fundamental level, the identification of a very large number of genes implicated in tumour phenomena (oncogenes and antioncogenes) and, therefore, identification of mutations in these genes, either in the tumours of affected patients (of both diagnostic and prognostic value), or in healthy patients at risk from a hereditary form of cancer (screening and prevention).

Finally, there is a field where the applications of molecular biology are so revolutionary that they have contributed as much to the popularization of molecular biology as to their medical application; this field is that of genetic fingerprinting applied to forensic medicine (paternity testing or criminal identification).

This book therefore proposes, after a simple yet exhaustive description of the principal techniques of molecular biology, to review the applications in the different disciplines written by a recognized expert in that discipline, without neglecting ethical questions or those that are raised by society in regards to current and future applications.

Jean-Louis Serre
Université de Versailles

List of Contributors

Catherine Boileau, CHU Paris, Île-de-France Ouest

Nevine Boutros, Hôpital Saint-Vincent-de-Paul, Paris

Anne Casetta, Hôtel Dieu, Paris

Denis Cointe, Hôpital Antoine-Béclère, Clamart

Véronique David, CHU Pontchaillou, Rennes

Annick Diolez, Institut national de la recherché agronomique, Versailles

Emmanuelle Girodon, CHU Henri-Mondor, Créteil

Liliane Keros, CHU Paris-Sud, Antoine- Béclère, Clamart

Éric Le Guern, CHU Pitié-Salpétrière, Paris

Jean-Paul Moisan, CHU de Nantes

Étienne Mornet, Université de Versailles, Saint-Quentin-en-Yvelines

Olivier Pascal, CHU de Nantes

Véronique Pingaud, CHU Henri-Mondor, Créteil

Serge Pissard, CHU Henri-Mondor, Créteil

Jean-Louis Serre, Université de Versailles, Saint-Quentin-en-Yvelines

Brigitte Simon-Bouy, Université de Versailles, Saint-Quentin-en-Yvelines

Dominique Stoppa-Lyonnet, Institut Curie, Paris

1 Techniques and tools in molecular biology used in genetic diagnoses

Jean-Louis Serre, *Université de Versailles*

This chapter is not meant to be an exhaustive survey. Its purpose is to make the reader familiar with molecular biology techniques and tools used for genetic diagnosis. Some specific techniques will be described in the chapter describing the diagnoses for which they are used.

1.1 Nucleic acids

DNA and RNA molecules are linear molecules made up of a succession of nucleotides. A nucleotide consists of a sugar (pentose with five carbons numbered 1′ to 5′) carrying a phosphate on its 5′ carbon and a nitrogenous base on its 1′ carbon (Figure 1.1). RNA contains a ribose while DNA contains a deoxyribose (no hydroxyl on the 2′ carbon, Figure 1.1).

Each deoxyribose of the DNA carries one of the four following bases: adenine (A), guanine (G), cytosine (C) and thyamine (T); for RNA, each ribose carries one of the following bases: adenine (A), guanine (G), cytosine (C) and uracil (U, acting in place of the thyamine). Another nitrogenous base is sometimes present but at a very low frequency.

Nucleotides are linked by a phosphodiester (sugar-phosphate) bond (Figure 1.1) between the 5′ phosphate of one nucleotide and the 3′ hydroxyl of the previous nucleotide. DNA and RNA are synthesized from the 5′ to the 3′ end in order to keep the 5′P and the 3′OH extremities in the same orientation (Figure 1.1).

In most organisms (plants, animals and some bacteria), the DNA molecules carrying the genetic information consist of two complementary and anti-parallel strands

Diagnostic techniques in genetics   Edited by Jean-Louis Serre
© 2006 John Wiley & Sons, Ltd

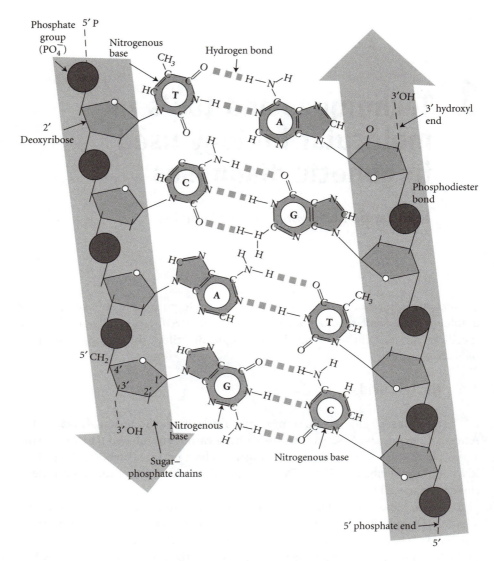

A DNA molecule consists of two strands held together by hydrogen bonds. Two hydrogen bonds in A/T base pairs and three in C/G pairs assuring the cohesion of the double helix also called a DNA duplex. Each strand consists of a nucleotide chain. The two strands are perfectly complementary. The two strands run antiparallel to each other.

Figure 1.1   Unwound structure of DNA (after J.-L. Rossignol (2000). *Génétique*. Dunod, Paris)

(Figure 1.1) linked by weak hydrogen bonds while RNA molecules are only single stranded. Bacteria and viruses are more diverse and have DNA or RNA genomes, single stranded or double stranded.

The two strands of a DNA molecule are complementary, meaning A is always associated with T and G is always associated with C, and vice versa. The complementary

---

Box 1.1   DNA molecules consist of two complementary strands

The two consecutive strands of the DNA always put a thymine in front of an adenine or a cytosine with a guanine, in such a way that knowing the sequence of one strand allows the deduction of the other one. This complementarity is a property used in natural mechanisms such as replication, transcription and translation but also in all *in vitro* technologies based on molecular hybridization, the capability of two single-stranded complementary sequences to associate and form a double-strand sequence.

---

nature of the two strands underlies the mechanism of replication: once the strands of a DNA molecule are separated, two new complementary strands are synthesized using each of the existing strands as a template.

The complementary nature of the two strands also underlies all molecular biology techniques using 'hybridization', meaning the ability of two complementary DNA strands to 'renature', or to 'hybridize' – the double-strand structure being thermodynamically more stable (lower free-energy level than the single-strand structure).

The two complementary strands of a DNA molecule are assembled in an antiparallel fashion, in an inverse orientation so that the 5′ and 3′ extremities are in opposite directions (Figure 1.1). This inversion of the 5′/3′ polarity of the strands is the basis for the helical structure of the DNA and for the different conformations of the binding sites for regulatory factors. It also provides information for the transcription machinery to distinguish between the coding and the non-coding strand at the level of the promoter. In addition, the translation machinery is affected by the 5′/3′ orientation of the messenger RNA, and knows how to identify useful sequences like the translation initiation codon. Therefore, when doing molecular biology, it is important to keep in mind that DNA fragments (genomic or not) which go through renaturation (hybridization) not only have complementary sequences but also opposite 5′/3′ polarity.

It is useful to remember that DNA synthesis, as well as RNA synthesis, consists of primer extension (RNA primer synthesized by a primase during replication) from its 3′OH extremity. Polymerization occurs in the 5′ to 3′ direction, using a DNA (or an

---

Box 1.2   The two strands of DNA molecules are also antiparallel

The 5′ phosphate extremity of a strand is always facing the 3′OH extremity of the other in a way that the sugar–phosphate chains are orientated in a reverse or antiparallel orientation. This reverse orientation of the two DNA strands constitutes a property used in the orientation of natural mechanisms of replication, transcription and translation, but also in most of the technologies based on molecular hybridization such as PCR, sequencing and cloning.

RNA) strand as a model, and to which the primer is hybridized in the opposite direction (3' to 5'). The duplex formed is complementary and has the opposite polarity.

Apart from the messenger RNAs that are transiently present during gene expression, other RNAs belong to the cellular machinery and constitute the 'machine tools' involved in gene expression and its regulation (ribosomal and transfer RNA for translation, nuclear RNA for RNA maturation and transport). RNA molecules are single stranded, although they often fold to form partial duplex domains.

1.2   The different types of genetic material studied

For diagnosis, molecular biologists mostly study and manipulate DNA, but RNA can also be studied, for example in virology where the genomes of many viruses are made of RNA.

1.2.1   DNA origins and types

Individuals are formed of clones of cells, each cell containing the constitutional genome, the totality of the genetic information present at conception in the cell (zygote) arising from the fusion of the gametes. The DNA from the constitutional genome, also called 'constitutional DNA', can be extracted from all healthy tissue biopsies. In most cases in man, it is extracted from lymphocytes after a blood test.

It is also useful to specify the origin of the DNA when it comes from a particular tissue sample, for example 'somatic' DNA as opposed to 'germinal' DNA or 'fetal' DNA (constitutional DNA extracted from fetal cells taken from an embryo *in vitro* or *in vivo*, amniotic cells taken from amniocentesis after 14 weeks of amenorrhoea, or trophoblastic cells obtained through biopsy after 11–12 weeks of amenorrhoea). Likewise, as DNA from tumour cells has local mutations compared to constitutional DNA (the tumourigenesis resulting from multiple mutations affecting a certain number of genes), DNA extracted from tumour tissue will be called 'tumour' as opposed to 'constitutional' DNA extracted from healthy somatic tissue.

1.2.2   RNA and cDNA

These are often the molecules of interest for molecular biologists within the framework of research or diagnostics, messenger RNA or viral RNA (see RT–PCR technique).

Isolation of messenger RNA is well understood in genes that code for peptide chains. These mRNAs, for the most part, carry a long polyA tail, by which they can be trapped within an affinity column containing a polyT carrying resin, binding in this column to the free oligodT (the synthesized form of a oligodeoxynucleotide from a thymidine mono-strand).

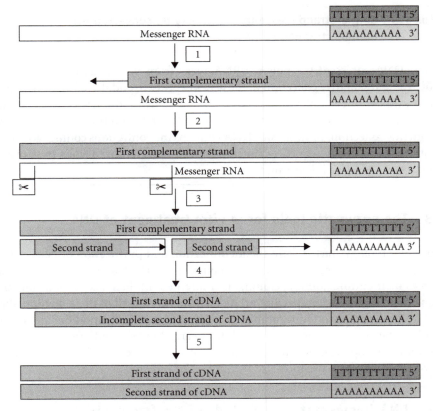

1. The reverse transcriptase synthesizes a strand of DNA complementary to an mRNA template strand, by lengthening the oligo-dT primer.
2. The messenger RNA strand is cut randomly and specifically by the action of RNAase H.
3. A DNA-polymerase l (*E. coli* or T4) is utilized to synthesize the second DNA strand, using the first as a template, by growth of the RNA primer and digestion of the RNA primers further down the strand.
4. The cDNA obtained are often not complete at the 5′ region of the mRNA.
5. Certain methods allow generation of a complete cDNA.

Figure 1.2 Synthesis of cDNA

Messenger RNA can be studied directly after purification (see Northern technique), but can also be converted into complementary DNA (cDNA), that is to say a double-stranded molecule of DNA made *in vitro* and without a 'natural' corresponding molecule. This is done using a reverse transcriptase that, in the presence of free triphosphate nucleotides, will start from an oligo-dT primer hybridized to the polyA tail and synthesize a primary strand of DNA that is complementary to that of the mRNA. After this, a DNA polymerase will synthesize a second strand of DNA using the first as a template (Figure 1.2). There are several methods for replacing the mRNA by the second DNA strand in order to obtain cDNA. The majority produce an incomplete cDNA where the 5′ part of the mRNA (which is generally non-coding) is missing, but methods have been developed to have a complete cDNA, an integral copy of the original mRNA sequence, in the form of double-stranded DNA.

> **Box 1.3   Single-strand messenger RNA may be converted into double-stranded cDNA**
>
> The cDNA copies of the polyA mRNAs vary from one tissue to another and form a representative collection of the genes expressed, and therefore transcribed into mRNA, in each of the tissues. It is possible using the RT–PCR technique (see below) to obtain in the form of cDNA, and then to amplify, only the mRNAs from this collection. In addition, cDNA contains the continuous coding sequence, with the non-coding sequences (introns) having been removed when the isolated mRNAs underwent nuclear maturation.

1.3   The enzymatic tools for *in vitro* treatment of DNA

Molecular biology started from the identification and purification of bacterial and viral enzymes allowing the *in vitro* manipulation of DNA and RNA nucleic acids, such as their cutting, joining, cloning, measurement of their size, establishing their sequence and identifying specific variations (mutations). All of these manipulations lead to the construction of recombinant genomes (genetically modified organisms) that have been used as tools in research for nearly 30 years, and more recently have had industrial applications in agronomy or pharmaceuticals. Table 1.1 shows a list of some of the enzymatic tools used in molecular biology.

1.4   DNA fragmentation and study of the fragments

1.4.1   DNA fragmentation

DNA fragmentation is currently one of the operations executed on DNA *in vitro*. Physical methods (pressure and ultrasound) perform well but work randomly, and molecular biology really started on the day that bacterial endonucleases were discovered. These enzymes, called restriction enzymes (because the protection they give against viral genome infections is 'restricted' to certain stocks), are endonucleases capable of cutting the DNA double helix, at a specific recognition and binding site (type II endonucleases), so that several identical DNA molecules are cut in the same manner, giving the same collection of fragments called 'restriction fragments'.

Certain restriction enzymes function in a way that cuts uniformly leaving fragments with blunt ends; others perform an overhanging cut creating fragments with single-stranded extremities. These extremities are often present in the form of complementary sequences because the restriction enzyme used in molecular biology recognizes a site in the DNA that constitutes a palindromic sequence. These single-strand complementary sequences are also called cohesive sequences because they allow the reassociation of the fragments through molecular hybridization of

Table 1.1 Name, function, operation and use of certain enzymes currently used in molecular biology

Name	Function (origin)	Operation and use
DNA pol 1	Polymerase (*E. coli*)	–DNA polymerase with 5'/3' activity –Lengthen 3'OH of a DNA or RNA primer from a DNA template –Exonuclease activity in reverse (3'/5') or forward (5'/3') direction (digestion nucleotide by nucleotide moving down the strand hybridized to the template)
Klenow fragment	Polymerase (*E. coli*)	–DNA pol 1, without any 5'/3' exonucleasic activity –Utilized in the synthesis of a second DNA strand from a DNA template: cDNA, labelled probe etc.
Taq pol	Polymerase (*Termophilus aquaticus*)	–DNA polymerase with 5'/3' activity by lengthening a DNA primer/DNA template –Thermostable, used for PCR cycling
Reverse transcriptase	Polymerase	–DNA polymerase 5'/3' activity by lengthening a DNA or RNA primer on an RNA template –Useful for the synthesis of cDNA and for RT–PCR
T4 DNA pol	Polymerase (*E. coli* infected by T4)	–DNA polymerase with 5'/3' activity and 3'/5' exonuclease activity, similar to the Klenow fragment –Preferable for certain protocols (directed mutagenesis)
Terminal transferase	Polymerase	–Lengthen DNA molecules by the incorporation of random nucleotides at the 3'OH end –Useful to generate polydT or dU tails carrying a marker
DNA ligase	Ligase (*E. coli*)	–Ensure the formation of a phosphodiester link between two 3'OH and 5'P ends –Useful for reattaching DNA fragments reassociated by hybridization of their 'sticky ends' and to form recombinant DNA
T4 DNA ligase	Ligase (*E. coli* infected by T4)	–Ensure the formation of a phosphodiester link between the 3'OH and 5'P ends of two fragments with sticky or blunt ends –Useful for reattaching the fragments to be cloned onto the blunt ends
Nuclease SI	Nuclease	–Digests exclusively single-stranded nucleic acids, double-stranded DNA/DNA or DNA/RNA are not affected

(continued)

Table 1.1   (*Continued*)

Name	Function (origin)	Operation and use
Exonuclease III	Nuclease	–Useful for generating blunt ends at the ends of the restriction fragments or to eliminate the unmatched DNA loops on a heteroduplex –Digests sequentially nucleic acid molecules from their 3′OH end (3′ phosphatase activity) –Useful in forming a simple strand of DNA from one of its extremities. The limited and combined action of exonuclease III and nuclease S1 allows deletions to be generated in a DNA molecule that has undergone a double-strand cut
RNase A	Endonuclease	–Exclusively digests single-strand RNA, no effect on duplexes –Useful for the purification of DNA, the detection of mismatches in a DNA/RNA duplex, nuisance factor during RNA polyA purification
RNase H	Endonuclease	–Digest RNA exclusively in DNA/RNA duplexes –Useful for the synthesis of cDNA
T4 polynucleotide kinase	Kinase (*E. coli* infected by T4)	–Transfers the γ phosphate of an ATP to the 5′OH end of a nucleotide –Useful for marking synthesized oligoprobes that are not 5′P but 5′OH
Alcaline phosphatase	Phosphatase	–Removes the 5′ phosphate of DNA or RNA –Useful for blocking ligase action at a given site
Eco RI	Endonuclease (*E. coli* strain R)	–The best-known 'restriction' enzyme and the first to be identified –Part of the type II class of restriction enzymes, making a cut in the DNA double helix at the level of its specific recognition and DNA binding site (see Table 1.2).

Box 1.4   Restriction nucleases are efficient tools to cut double-helical DNA into specific fragments

Each restriction enzyme cleaves (digests) the double-helical DNA at a specific site called the restriction site (digestion site), in such a manner that the DNA molecules are cleaved (digested) in exactly the same manner into a specific series of restriction fragments (digestion fragments).

Table 1.2 Examples of Type II restriction enzymes

Name	Restriction site (/ indicates cut)	Site of the cut in the 5′/3′ direction Consequences and Remarks
Eco RI	5′ G/AATTC 3′ 3′ CTTAA/G 5′	Overhanging cut, sticky ends 5′AATT
Msp I	5′ C/CGG 3′ 3′ GGC/C 5′	Overhanging cut, cohesive extremities 5′CG
Hae III	5′ GG/CC 3′ 3′ CC/GG 5′	Blunt cut, between G and C *Note*: the site differs from that of *Msp* I because of the 5′/3′ orientation
Sma I	5′ CCC/GGG 3′ 3′ GGG/CCC 5′	Blunt cut, the *Sma* I site includes an *Msp* I or *Hpa* II site, the *Sma* I site is more rare than the *Msp* I (longer palindrome), the restriction fragments of *Sma* I are, on average, longer than those of *Msp* I
Apa I	5′ GGGCC/C 3′ 3′ C/CCGGG 5′	Overhanging cut, sticky ends GGCC3′, includes a Hae III site, gives longer fragments, different from the Sma I site due to the 5′/3′ orientation
Ava II	5′ G/G(A or T)CC 3′ 3′ CC(T or A)G/G 5′	Imperfect palindrome, overlapping cut, the 5′G(T or A)C extremity is not obligatorily sticky
Bgl I	5′ GCCNNNN/NGGC 3′ 3′ CGGN/NNNNCCG 5′	Imperfect palindrome with any central bases, NNN3′ extremities are rarely sticky
Hpa II	5′ C/CGG 3′ 3′ GGC/C 5′	Overlapping cut, sticky ends 5′CG, same site as Msp I, but sensitive to methylation of the second cytosine (if methylation, no cut) useful for showing the DNA methylation status at this site

their extremities, an operation that is the basis of the construction of recombinant DNA.

The specific site of connection and cleavage, called the 'restriction site' differs from one enzyme to another (Table 1.2). It consists of a palindromic sequence (complementary sequence, identical in both directions) of between four and eight base pairs. The palindrome may either be perfect or imperfect and the cut may be either blunt or overhanging (Table 1.2). In the latter case, the 'restriction fragments' each possess the same single-strand end (Figure 1.3). This end is called 'sticky' because the strands with the same ends can assemble themselves by renaturation of these extremities.

1.4.2 Separation of DNA fragments by electrophoresis and membrane transfer

The goal of DNA fragmentation research is often the study of one or more particular fragments, which are isolated by separating them from other fragments, and are then identified.

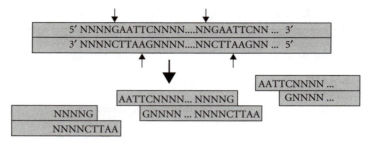

The specific binding site of *Eco* RI is 5'GAATTC3'; the cut is situated after the G, freeing the 'restriction' fragments.

Figure 1.3   Action of Eco RI endonuclease

Electrophoresis

This operation consists of separating the molecules in a mixture (in this case the restriction fragments) by placing them in an electric field at one end of a polymer gel (agarose or polyacrylamide) which, via pores in the gel, acts as a sieve. The negatively charged DNA fragments (Figure 1.4) migrate in the direction of the positive pole with longer molecules migrating more slowly as they are slowed down by the filtration action of the gel. When the electrophoretic migration is stopped, the fragments are separated with the longest ones remaining closest to the entry point of the mixture and the shortest found furthest away if they have not left the gel entirely. It is possible to play with the timing of the electrophoresis and the size of the pores in order to provide the best resolution (separation) among the fragments to be separated and studied.

Results are visualized (Figure 1.4) by the addition of ethidium bromide (either in the gel or after migration depending on the situation), a dye with high affinity for DNA that will fluoresce orange when exposed to UV light. By placing the electrophoresis gel in a UV transilluminator the separation of the fragments can be seen. When the initial mixture is made up of a large number of different fragments (for example the enzymatic digestion of genomic DNA), transillumination by UV will just show a continuous streak, but if the mixture is made up of a limited number of fragment types it is possible to count and measure the size of the fragments by comparing the distance of migration of standard fragments in a parallel well. A range of fragment standards of known size is used in electrophoresis within the agarose gel (for example a phage λ genome, digested by Hind III enzyme, or phage φX genome 174 digested by Hae III). An abacus (a graph which shows the size of the standard fragments and their migration distance) allows precise calculation of the size of all of the identified fragments in the gel, or later on a Southern blot (see below).

There are different methods of electrophoresis that are specifically adapted to the size of the molecules to be separated and/or identified (types of gel polymer and electric field). Table 1.3 shows, in a simplified form, the range of uses of several kinds of electrophoresis.

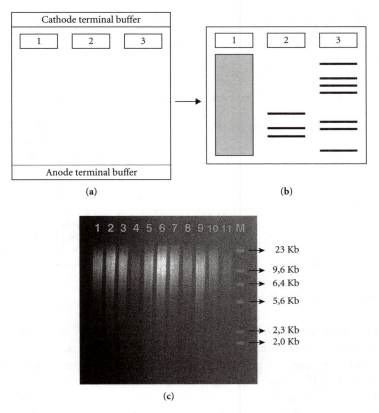

(a)   (b)

(c)

(a) The wells are loaded with samples of restriction fragments from genomic DNA (1), a mixture of three plasmids of differing sizes and linearized at a specific site (2), and a range of standard phage λ restriction fragments digested by *Hind* III (3), making respectively 23, 9.6, 6.4, 5.6, 2.3, 2.0, 0.5 and 0.125 kb.

(b) The fluorescence of the ethidium bromide allows visualizing of the genomic fragment smear (see (c)), the separated plasmids and the range of standards (see (c); the smallest fragment at 0.125 kb is always closest, if not past, the anode terminal).

(c) Gel showing the smears and the molecular weight marker (M).

Figure 1.4   The principal of electrophoresis on agarose gels

---

Box 1.5   Gel electrophoresis separates DNA restriction fragments of different sizes

After the application of a restriction enzyme, the digested fragments can be subjected to electrophoresis, a process that allows the separation of fragments based on their size within a gel subjected to an electrical field. The location of the fragments of interest after electrophoresis separation can be direct if the fragments are individual or labelled, or indirect if the fragments were transferred from another support (Southern blot), so as to be recognized by molecular hybridization with a labelled probe within the complementary sequence or the target sequence.

Table 1.3 Types of electrophoresis as a function of the size of the DNA fragments to be separated

Support polymer and concentration	Type of electric field	Size of separable fragments and range of utilization
Agarose between 0.6 and 1.5 %	Unidirectional	From 0.2 to 20 kb used for genomic fragments
Polyacrylamide 4 to 11%	Unidirectional	From 10 to 800 kb with a possible resolution of base pairs when used with small fragments (used after PCR or for sequencing)
Agarose 1%	Pulse field (multidirectional and alternating)	Large fragments prepared by extraction and digestion arranged from 20 to 150 kb Field inversion gel electrophoresis (FIGE) from 100 to 2000 kb or more Pulse field gel electrophoresis (PFGE) used for YACs or small chromosomes

Membrane transfer (Southern blot)

Most of the time, the fragments of interest are not visible in the gel because they are mixed with too many other restriction fragments (Figure 1.4, well number 1 and photo). Moreover, they are not easily accessible with a reactive inserted in the gel. Finally, the gel itself is fragile and is not easily manipulated.

This is why, in these situations, the DNA fragments separated by electrophoresis are transferred to a nylon membrane (Figure 1.5), a solid support where they are made easily accessible to analytical techniques and identification. As this identification technique (hybridization with a probe) requires single-stranded fragments, the gel is subjected to a sodium bath in order to denature the DNA fragments, a blotter sheet soaked in buffering solution is then placed on top of the gel followed by a stack of wadding on top. This, by means of capillary action, causes the vertical transfer of the fragments in the gel onto the membrane without modifying their respective distances resulting from the electrophoresis. The fragments are now single-stranded, accessible and attached in an irreversible fashion onto a solid support (some covalent bonds can be stimulated by UV on certain types of membranes). This technique, invented by Ed Southern, carries his name (*Southern blot*: Southern transfer; Figure 1.5).

By analogy, the transfer of RNA separated by electrophoresis onto a membrane is called a *Northern blot* and the transfer of proteins separated by electrophoresis is called a *Western blot*.

1.5 Selective amplification of a nucleotide sequence

Several methods have been developed to amplify selectively either a DNA or RNA sequence. The only purpose of some of these methods is to provide a large quantity

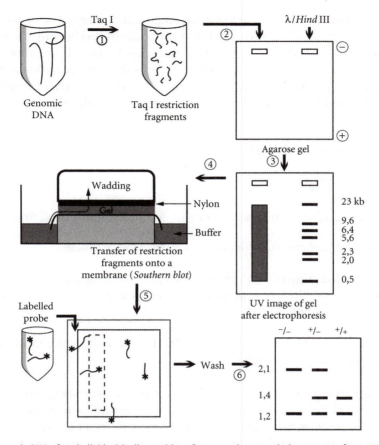

1. The genomic DNA of an individual is digested into fragments by a restriction enzyme, for example Taq1.
2. The fragments are separated by electrophoresis.
3. The electrophoresis is visualized (see Figure 1.4).
4. The fragments are transferred onto the nylon membrane: Southern blot.
5. The membrane is hybridized to a labelled probe, which is specific to the fragments to be identified.
6. After washing, the revealing of the labelling allows the position of the fragments hybridized to the probe to be located; the diagram shows the three possible images for the studied individual with an RFLP specific probe (see Section 1.9.8).

Figure 1.5   Realization, finishing and analysis of a Southern blot

of the sequence of interest (qualitative methods), while some other methods are set up to measure the amount this sequence (quantitative methods).

All of the amplification techniques are second-generation techniques that assume that the fragment of interest has already been cloned and sequenced (see below), at least partially at the extremities. In this case, only a small amount of genomic DNA (or plasmid DNA) containing the fragment of interest, or a few RNA molecules in a mix, are necessary to amplify the fragment *in vitro* and obtain a large enough quantity to perform the study, the digestion, separation, transfer and first-generation analysis techniques then becoming useless.

The most famous and most used specific amplification method is the polymerase chain reaction (PCR) developed in 1985 by Kary Mullis (Nobel Prize for Chemistry in 1993). It revolutionized molecular biology and its applications by offering a simple, economical and fast (a few hours) routine whereas the first-generation techniques were laborious, expensive and time consuming (several days). Other amplification methods less widely used except for RNA amplification are described later.

1.5.1   DNA sequence amplification by PCR

PCR framework (Figure 1.6)

The DNA containing the fragment of interest is pipetted into a small tube with:

- the four triphosphate nucleotides necessary for DNA synthesis

- single-stranded primers (of 15 to 30 bases) corresponding to the 5′ extremities of the two strands of the fragment to be amplified, labelled a/+ for the + strand and b/− for the − strand

- a few microlitres of Taq polymerase (Table 1.1) or another thermostable DNA polymerase in an ad hoc buffer.

The primers are synthetic oligonucleotides, which requires knowledge of the sequence of the 5′ ends of the + and − strands of the DNA fragment to be amplified.

The tubes are then closed and submitted to three different temperatures during a determined amount of time, this making up one cycle (Figure 1.6) that will be repeated n times (n varies between 10 and 50) depending on the purpose of the experiment. As the cycle is repeated n times, the number of copies of the fragment obtained is theoretically 2^n (see the limitations below).

Note 1 PCR reactions are feasible thanks to the discovery of the thermostable *Taq pol.* Otherwise, it would be necessary to add some polymerase in each tube at each cycle, a process that would be time consuming, expensive and a cause of contamination (see below).

Note 2 The n PCR cycles are generally preceded by a denaturation step of several minutes because the starting material generally consists of large genomic DNA duplexes; for each of the n cycles, a 30 to 90 s denaturation step is sufficient because the DNA molecules to denature (amplicons) are short.

Note 3 The PCR quality is assured by the combination of several technologies in physics and informatics. The tubes submitted to PCR are inserted in a thermal block in which the temperature can increase or decrease at a rate of 1 to

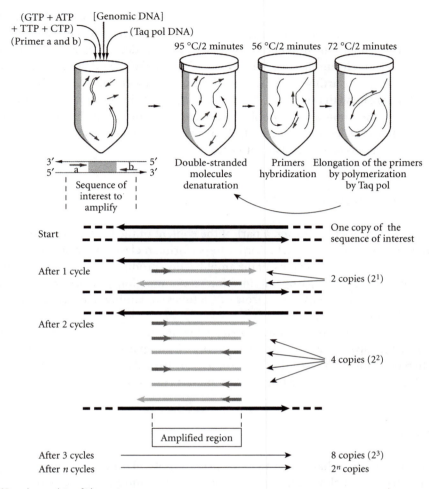

A PCR cycle consists of three steps:

Step 1: denaturation of the DNA duplex by heating at 95°C for 30 to 90 s.

Step 2: hybridization of the a/+ and b/− primers with the − and + strands for 30 to 60 s, at a temperature allowing their specific hybridization with the 5′ ends of the target sequence.

Step 3: elongation of the primers: the tube is kept for 30 to 120 s at 72°C, the temperature corresponding to the optimum activity of the Taq pol which elongates the primers and synthesizes a new strand from the − and + templates, which doubles the quantity of the target fragment.

Figure 1.6 The principle of the PCR

2 °/s; the block temperature is measured by sensors and a computer program calculates the temperature inside the tube which is in contact with the block; another computer program controls the increase and the decrease of the temperature and the thermal inertia in order to avoid any overheating or overcooling and to start the countdown when the tubes reach the selected temperature.

Note 4 At the end of each cycle, the strands – newly synthesized using the genomic template – are only terminated at one of their extremities corresponding to the primer. The number of copies with one border increases by one at each cycle, their number reaching n after n cycles. On the other hand, after each cycle and starting from the second cycle, the new strands synthesized from a template with one or two borders will definitely be terminated at both ends and their number will double and increase exponentially in such a way that after n cycles, most of the DNA in the tube would correspond to amplified fragments terminated at both ends.

Control and analysis of the amplification products

Once the PCR reaction is over, a part of the content of the tube is deposited on an agarose or polyacrylamide gel in order to perform an electrophoresis to check the presence and/or the conformity of the amplification products. It also enables a check to be made for the absence of amplification (contaminations, see below) in the wells where the 'blanks' or negative controls – PCR tubes containing all the reagents except the DNA – were loaded.

The control gel is itself an analysis step when the size of the fragment of interest is verified (analysis of a mutation characterized by a small deletion or insertion or a motif repeat). In other cases, the PCR products are treated and then analysed on a new gel; this is the case when the fragment of interest contains an RFLP site (see below).

Dot blot and slot blot

If the PCR products have to be analysed using a molecular hybridization technique with a probe (see below), they have to be transferred onto a membrane in the same way that restriction fragments are transferred to perform *Southern blot*. However, it is useless to perform an electrophoresis because the tube content is supposed to be homogenous and should only contains copies of the fragment of interest, except for a few genomic DNA molecules. Therefore the PCR products just have to be denatured and deposited on a membrane called a *dot blot* (the term *slot blot* is sometimes used to refer to the resulting slide that shows slots like a slot machine).

Limitation of the yield

The theoretical number of 2^n copies is seldom reached: first because the fragments present in the tube are not all necessarily used as a template, above all at the end of the PCR, also because some inhibitors like pyrophosphate are produced, and finally because at the end of the PCR, the amplified fragments, particularly if they are short, compete with the primers whose concentration is decreased. The yield is likely to give 1.85^n copies in optimal conditions.

Risks of errors: contamination, non-specific amplification or infidelity

Contamination is a permanent danger due to the presence in the atmosphere, particularly in a laboratory, of exogenous DNA that can contaminate the studied sample and falsify the result of the analysis and, therefore, the diagnosis. Therefore, a great deal of care must be taken in filling the PCR tubes and all PCR routines should use negative controls, tubes containing all the reagents apart from the DNA and whose content should not give any visible band on the control gel in the absence of a contamination.

Non-specific amplification results from the hybridization, by homology, of the primers elsewhere than at the extremities of the fragment of interest; this is why it is necessary to choose a pair of primers long enough, purified (the chemical synthesis is not error free), with identical or close annealing temperatures, which are as high as possible.

Copy infidelity results from the fact that, *in vitro*, *Taq pol* has a relatively high rate of incorporation errors (10^{-4}). A mis-incorporation generates a mutated fragment, and the earlier the mis-incorporation occurs, the higher the proportion of mutated fragments among the amplimers because the amplification process will multiply the mutated 'strands'. Such a rate is admissible if the criterion for the analysis of the fragment of interest, after the PCR, is its length, but it can raise some problems, notably in the case where PCR products are sub-cloned for sequencing or in order to perform expression studies (some sub-clones can carry a mutated sequence instead of the original sequence).

Box 1.6(a)   Specific DNA sequences may be amplified by PCR and RT—PCR

PCR is a technology that allows selective amplification of a specific DNA sequence. This technology relies on thermoresistant polymerases, a necessary condition to go through the denaturation steps at 95°C that occur at the beginning of each amplification cycle. This step is followed by an annealing step during which the primers (synthetic oligonucleotides) anneal (molecular hybridization) with each strand of DNA, at one of the two extremities of the sequence of interest. The third step of elongation allows the synthesis of the new strand by the polymerase by elongation of the primer using the DNA strand as a template. After one cycle, the number of copies of the target sequence is doubled, after two cycles, it is multiplied by four, after n cycles, it is multiplied by about $2n$.

Hot start PCR

The simple hot start consists of adding *Taq pol* after the initial long step of denaturation (see earlier note), in order to prevent the polymerase from spending several minutes at 95°C (its thermostability, perfect at 85°C, is not so great at 95°C, especially over a long period of time).

However, opening the tubes before the *n* cycles to add the enzyme to the tubes increases the risks of contamination and of parasitic amplification, as well as the extra time necessary to distribute the enzyme into the tubes. Even if this simple hot start can still be done, the biotechnology company have now developed several molecular systems that, once dispatched in a tube, require several minutes of incubation at 95°C before release of a native and active *Taq pol*.

1.5.2   RNA amplification as cDNA by RT–PCR

RT–PCR is a PCR designed to amplify selectively as a double-stranded DNA fragment, a specific RNA sequence present within a mix of polyA-mRNA or total RNA, isolated from a tissue expressing the gene of interest. This is also a second-generation technique that allows obtaining the cDNA from a specific mRNA, without having to prepare or to sort a cDNA library, but that requires that the sequence of interest is already known in order to define the primers.

The RNA mix is incubated with a DNA primer, called b, complementary to the 3′ terminal sequence of the sequence of interest and with a DNA primer, called a, corresponding to its 5′ terminal sequence, in the presence of the four triphosphate nucleotides, reverse transcriptase and *Taq pol*. At 37°C, only the reverse transcriptase is active and synthesizes a (−) strand complementary to the mRNA starting from the b primers that are hybridized to the 3′ end of the mRNAs. The tubes are then transferred to a PCR plate for *n* cycles; the reverse transcriptase is denaturized and only *Taq pol*

Box 1.6(b)   Specific RNA sequences may be converted into double-stranded DNA and amplified by RT–PCR

RT–PCR consists of the amplification of a messenger RNA in a selective way and as double-stranded DNA. For this purpose, an antisense strand of DNA is made by a reverse transcriptase starting from an antisense primer specific to the 3′ extremity of the mRNA of interest. The second strand of DNA is then synthesized by a classic polymerase, starting from a primer specific to the 5′ extremity of the antisense DNA strand. From the moment when the two DNA strands are available, a classic PCR using the same primers or more internal primers, will amplify the target sequence.

is active. During the first cycle, it uses a primers to make the (+) strands using as a template the (−) strand of DNA synthesized by the reverse transcriptase (*Taq pol* can only use DNA templates), re-establishing an equal quantity of (−) and (+) strands of the sequence of interest; the latter then starts to be amplified at the second PCR cycle. Nowadays, the Tth polymerase is available (a thermostable polymerase from *Thermus thermophilus*) that makes the use of the reverse transcriptase unnecessary because it can use both DNA and RNA templates.

1.5.3   Quantitative PCR methods

The purpose of a diagnosis is sometimes quantitative; it consists of the quantity of a sequence of interest contained (either mixed or integrated) within a certain mass or volume of biological material; for example, the number of copies of a transgene within a transformed genome, or the percentage of genetically modified seeds within a possibly contaminated batch of seeds. The dosage of a sequence of interest using PCR consists of the estimation of the number of copies of this sequence within the tube at the beginning of the process. A simple rule of three is sufficient to estimate the total quantity of the sequence of interest contained in the studied biological or genomic material knowing the volume tested by PCR. Two quantitative PCR strategies can be defined depending on whether the dosage is terminal – meaning performed at the end of the PCR on the final amplimers, or real time, during the PCR steps. These two types of strategies are illustrated by competitive PCR and TaqMan PCR® respectively.

Competitive PCR

This method allows estimation of the quantity of the sequence of interest or 'target sequence' within a biological sample containing a known quantity of DNA. The principle consists of the co-amplification by PCR of increasing amounts of the target sequence together with a given amount of internal standard. This standard generally corresponds to the target sequence sub-cloned into a plasmid and deleted or increased by about 20 base pairs. The amplification primers are the same and the final quantities of amplimers, easy to separate on a polyacrylamide gel thanks to the different length of the two sequences, will only depend on the initial quantities of the target sequence and of the standard sequence respectively. The quantity of the target sequence within the sample can then be estimated knowing the quantity of co-amplified standard and comparing the lanes on the gel where the quantities of amplimers are the same (Figure 1.7). A simple rule of three will allow the determination of the total quantity of sequence of interest in the total volume of the studied biological sample or transgenic species genome. Of course, this technique requires a good standardization technique

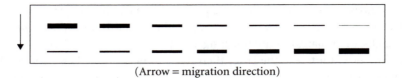

(Arrow = migration direction)

The amplimers of the standard sequence (cloned sequence equivalent to the target sequence, but with a central insertion of 20 bp) are longer than those of the target sequence to be quantified, and migrate less far. They correspond to the quantities present in the top band and are obtained in each tube from the same quantity of standard sequence.

The amplimers of the target sequence correspond to the quantities present in the lower band and are obtained by using increasing quantities in each tube.

The third lane corresponds to the tube where the quantities of standard and target amplimers are the same because the quantity of amplimers synthesized as a competitor is the same, which makes it possible to deduce the quantity of the sequence of interest in the tube, and in the DNA sample on which the PCR has been performed.

Figure 1.7   Control gel of a competitive PCR

in order to define the quantities of standard likely to correspond to the amount of target sequence.

PCR associated with TaqMan® chemistry

This quantitative method makes it possible to follow in real time the selective amplification of a sequence and therefore to deduce the number of sequences of interest contained in the tube knowing the number of cycles already performed.

One problem raised by real-time quantitative PCR comes from the fact that the detection of the amplimers supposes a minimal quantity that is only reached after a few cycles, each detection method having its own threshold. From the moment the amplimers are detectable, it is theoretically possible to measure their exponential increase during a few cycles before this increase stops at the end of the PCR, when inhibition (see above) or competition phenomena seriously limit the efficiency of the cycles. The part of the exponential phase usable for the real-time dosage will be limited to a few cycles and defines the window of detection.

A second problem raised by real-time quantitative PCR concerns the detection method. In some special circumstances (see Down's syndrome molecular diagnosis), it is possible to interrupt the PCR at one point of the window of detection, to compare the quantities of amplimers for the sequence of interest (chromosome 21 markers) with the one of a co-amplified control sequence (markers on other chromosomes). Indeed, in this special case, the quantitative ratio between the sequence of interest and the control sequence is not too unbalanced and can take previously known values, depending on whether or not there is a trisomy. In most circumstances, the sequence of interest is seriously in the minority compared with

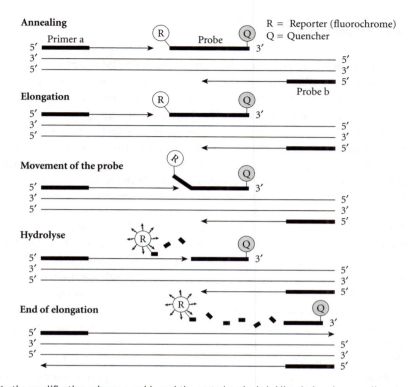

Figure 1.8 Principle of TaqMan® quantitative PCR

Step 1: the amplification primers a and b, and the central probe hybridize during the annealing step.
Step 2: during the elongation of the a primer, the Taq polymerase reaches the hybridization site of the central probe.
Step 3: the polymerase moves to the extremity of the probe.
Step 4: the polymerase hydrolyses the probe while moving forward, separating the nucleotides carrying the fluorochome (reporter) from the quencher.
Step 5: the polymerase terminates the synthesis of the new strand.
Result: at each doubling of the sequence of interest, a nucleotide carrying the fluorochrome, capable of emitting one unit of fluorescent signal, is freed in the buffer; the quantity of emitted signal gives a measure of half of the amplimers synthesized in the first cycle from the target sequence.

any other control sequence in a way that their co-amplification would lead to a saturation of control sequence even before the detection level for the target sequence can be reached. It is therefore necessary to quantify precisely and continuously the increase in the number of amplimers for the sequence of interest, without any control sequence co-amplification. This is when the TaqMan® chemistry is useful.

TaqMan® PCR (Figure 1.8) consists of a PCR on a target sequence, adding to the two amplification primers a central probe that can hybridize with one of the two DNA strands. This probe has at its 5' end a fluorochrome R whose emission is

---

Box 1.7   Quantitative PCR

The goal of PCR quantitative methods is to estimate the quantity of a target sequence within a population of molecules. The density of this sequence, generally very low, justifies the use of the PCR amplification technique. The target sequence can be quantified by co-amplification of a standard made up of various quantities of another sequence accepting the same primers (competitive PCR). It is also possible to follow in real time the exponential amplification of the target sequence using an ad hoc display (TaqMan® chemistry), and to calculate by extrapolation the initial number of copies of the target sequence in the tested sample.

---

quenched by the action of a molecule Q ('Quencher'), due to the transfer of energy resulting from the proximity of R and Q (Förster effect). During the hybridization step of each PCR cycle, the central probe can hybridize to one of the two strands. During the elongation step of one of the primers, the Taq polymerase will then reach the central probe, move it and hydrolyse it thanks to its $5'/3'$ exonuclease activity. This hydrolysis will free the nucleotide carrying R which, once released from Q, will emit a fluorescent signal. Therefore, at each amplification of the sequence of interest, an R molecule is freed in a way that the measure of the fluorescence emitted from the PCR tube, in real time, is a stoichiometric measure of half of the amplimers already obtained. Considering the number of cycles already performed, it is possible to deduce the initial number of copies of the target sequence. The measure of the fluorescence is performed using a display *in situ* in the amplifier, linked to a computer which graphically traces the increase in quantity of the amplimers and calculates the starting quantity (see Section 5.3, GMO tracing)

1.5.4   *RNA or DNA isothermic NASBA® amplification*

Isothermic amplification nucleic acid sequence based amplification (NASBA) allows exponential amplification of the sequence of interest, much faster than PCR and without needing a specialized automated system to deal with the programmed temperature cycles. Although it allows the amplification of both RNA and DNA, NASBA isothermic amplification has essentially been developed to amplify RNA selectively, notably in clinical virology.

The principle consists of the incubation of the biological material, assumed to contain the RNA of interest, in the presence of three enzymes, a reverse transcriptase (RT) capable of using an RNA or a DNA template, RnaseH (RH) and T7 RNA polymerase (T7P) and two primers labelled a(+) and b(−). The b(−) primer is complementary to the 3' end of the RNA of interest and carries at its 5' end a promoter sequence for T7 RNA polymerase; this promoter is required for transcription initiation in one of the steps of the amplification cycles. The tube is at a fixed and

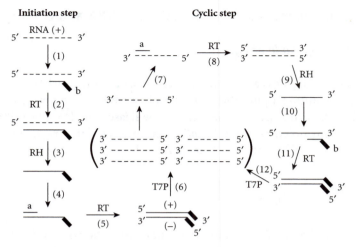

Figure 1.9 NASBA® isothermic amplification

Initiation step

Step 1: the b primer hybridizes at the extremity of the RNA target sequence.
Step 2: the RT elongates the b primer and synthesizes a DNA (−) strand carrying a 5′ extremity corresponding to the T7P promoter.
Step 3: the RH frees the (−) strand of the sequence of interest by hydrolysing the RNA from an RNA–DNA duplex.
Step 4: the a primer hybridizes with the (−) DNA strand of the free sequence of interest.
Step 5: the RT elongates the a primer and forms the double-stranded cDNA of the target carrying a T7P promoter.
Step 6: the T7P uses the promoting sequence to transcribe many antisense RNAs (as a (−) strand).

Cyclic step
Step 7: the a primer hybridizes at the extremity of the (−) strand of the RNA target sequence.
Step 8: the RT performs the elongation of the a primer and synthesizes the DNA (+) strand complementary to each RNA (−) strand.
Step 9: the RH hydrolyses the RNA from the RNA–DNA duplex and frees the (+) strand of the target sequence.
Step 10: the b primer hybridizes with the DNA (+) strand of the freed target sequence.
Step 11: the RT performs the elongation of the b primer and of the (+) strand for the promoting sequence for T7P brought by the b primer; a double-stranded cDNA carrying a promoter for the T7P is formed.
Step 12: the T7P uses its promoting sequence to transcribe a large number of copies of the target RNA as (−) strands.

the synthesized RNAs are therefore recycled for a new six-step cycle, without a need for the primers and the enzymes to act synchronously.

Result: at each cycle, the sequence of interest is not copied in two sets but in as many copies as the T7P can synthesize (around 1000). At the end of the reaction, the tube contains the sequence of interest amplified both as cDNA and as about 100 times more RNA (−) strands.

optimal temperature of 41°C at which the primers can hybridized and the enzymes are active. The amplification starts by an initiation for six steps followed by a series of a few asynchronous cycles, also in six steps (Figure 1.9), allowing the multiplication of the sequence of interest by several thousands.

1.6   DNA fragment ligation: recombinant DNA and cloning

1.6.1   Operating mode of ligases

DNA ligases are enzymes capable of rejoining DNA fragments when they are end to end, by reconstituting a phosphodiester bond between a 3′OH extremity and a 5′ P extremity (there are also RNA ligases). Some ligases (T4 DNA ligase) can rejoin blunt-end fragments, others like *E. coli*'s can only act if the fragments have been previously renatured by virtue of their sticky ends.

1.6.2   Recombinant DNA

The action of the ligases, the ligation, allows the construction of recombinant DNA by joining DNA fragments coming from different species as long as they have sticky ends. The fragments of different origins are pooled in a single tube named the ligation tube, where the ligase will combine them in various combinations; a selection screen should then be set up to isolate the desired construct. The construction of recombinant DNA is at the basis of DNA cloning and the construction of genetically modified organisms.

1.6.3   DNA cloning

Cloning a DNA fragment consists of the insertion of this fragment in a cellular host in a way that this host will replicate the exogenous DNA fragment as it will replicate its own genome. In other words, cloning a DNA fragment is the same as creating a genetically modified organism that acquired this fragment in its genome.

However, behind this general definition, there are a large variety of modalities:

- *depending on the nature of the cellular host*, micro-organism (bacteria or yeast), plant or animal;

- *depending on the localization of the cloned fragment*, inserted into the genetic material (bacterial chromoneme or eukaryotic chromosome) or simply added to it (plasmid or artificial chromosome);

- *depending on the transformation mode of the genetically modified cells*, insertion of naked DNA or via a vector;

- *depending on the preparation mode of the fragment of interest*, mixed with many other genomic fragments or PCR amplified. In the first case, the cloning of all the fragments leads to the construction of a library (Figure 1.10) that will have to be screened to identify and isolate the clone carrying the fragment of interest; in

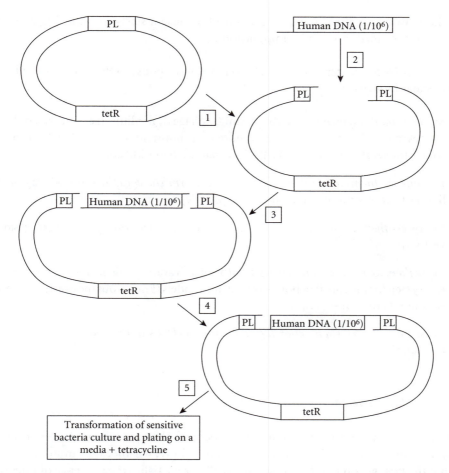

1. A bacterial plasmid carrying an antibiotic resistance gene, for example tetracycline resistance (tetR), is linearized at the E site of the polylinker (PL), a sequence used as an insertion site and containing a unique restriction site; the 5′ ends are dephosphorylated with an alkaline phosphatase, avoiding a later action of the ligase that would circularize the plasmid.
2. The human genomic DNA has been digested by an E enzyme giving sticky ends (or such extremities were added if it is cDNA), and is mixed with the plasmid (the figure represents the destiny of one fragment out of a million different fragments).
3. The ligation mix allows an eventual renaturation between type E cohesive ends and the formation of recombinant plasmids.
4. The ligase links covalently the 5′P ends of the fragment with the 3′OH extremities of the plasmid, creating an insert into the plasmid.
5. Tetracycline sensitive bacteria are transformed by penetration of the plasmidic DNA, cultured and plated in the presence of tetracycline; the clones that develop have acquired a plasmid that confers resistance and contain an insert, a necessary condition for the plasmid to circularize and replicate.

Each clone isolated from the plate comes from a single cell, its cells all contain the same insert; the corresponding human DNA fragment has therefore been isolated and amplified; it has been cloned. Altogether, the clones obtained this way constitute a genomic library or a cDNA library depending on the origin of the cloned fragments; it has to be screened to isolate the fragment of interest.

Figure 1.10   Construction of human DNA library in a bacterial plasmid

the second case, the cloning is more direct because it starts with the fragment of interest purified through its amplification.

Many host/vector systems have been developed in response to the different needs of molecular biologists:

- *to clone small fragments* (few kb: plasmids, phages), medium (several tens of kb : phages, cosmids) or large (several hundreds or thousands of kb: artificial chromosomes for yeast/ YAC, bacteria / BAC or mammalian / MAC);

- *to sequence them or use them in order to perform cartography* (ad hoc fragment library to walk or jump along the chromosome)

- *to express then* in vitro *or in vivo*, in bacterial or eukaryote systems (expression vectors);

- *to use them as a directed mutagenesis tool* by inactivation or allelic replacement, necessary step for the construction of transgenic models (knock-out mouse, transgenic Drosophila or mouse . . .);

- *to use them to perform functional studies*, protein-DNA or protein–protein interaction studies.

1.6.4   Cloning vectors

Plasmids are the most used cloning tools because they are easy to purify, to manipulate *in vitro* and to introduce into cells in culture, whether they multiply apart from the host genome or they integrate totally or partially (the sequence of interest carried by the plasmid) into the genome. Nowadays, very sophicated plasmids are available, with the cloning site for the sequence of interest surrounded by efficient sequences assigned to specific purposes (Figure 1.11). It is therefore possible to clone a sequence and to express the protein it codes for in most cell types in order to perform sequencing, functional analysis and production.

The sequences cloned in such plasmids have two possible origins, they can be prepared by PCR starting from a known original sequence kept in small quantities, or they can correspond to cDNA prepared from mRNA expressed in a tissue. In both cases, a different solution is found to make sure the open reading frame is kept between the initiation codon contained in the N-ter sequence (see Figure 1.11(b)) and the STOP codon for the fusion protein including the product of the cloned sequence (see Figure 1.11(g)).

In the first case, the amplimers inserted into the plasmid have to be in frame with the other sequences of the plasmid (see Figure 1.11(g)); knowing that all the amplimers are identical and are inserted at the same position into the plasmid, the PCR

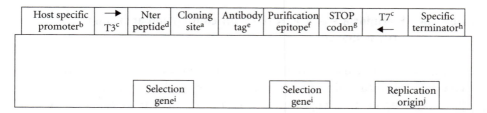

Host specific promoter[b]	→ T3[c]	Nter peptide[d]	Cloning site[a]	Antibody tag[e]	Purification epitope[f]	STOP codon[g]	T7[c] ←	Specific terminator[h]
		Selection gene[i]			Selection gene[i]		Replication origin[j]	

Plasmids are now constructs aligning efficient sequences for specific purposes. The described sequences are not necessarily present simultaneously.

(a) Cloning site: it allows the plasmid to close very efficiently on the sequence of interest, PCR fragment or cDNA (see below),

(b) Promoter allowing the expression of the cloned sequence in a cellular host, bacteria (for example: Arabinose operon promoter, that can be activated by arabinose), yeast (for example, GAL1 gene promoter that can be regulated by the product of the GAL4 gene), mammalian (for example, strong cytomegalovirus promoter),

(c) T3 or T7 sequence: promoter for T3 or T7 phages, allowing the transcription in vitro of the sense or antisense strand; these sequences (or others) can also serve as hybridization sites for elongation primers for the sequencing of the insert.

(d) Sequence allowing the fusion of the peptide chain encoded by the cloned DNA to a terminus sequence conferring diverse properties (for example the secretion in the bacterial periplasmic region or increasing its solubility); this sequence is generally terminated by a site for a protease, that allows the cleavage of the fusion protein after purification in order to get the original N-ter extremity of the cloned sequence.

(e) Monoclonal antibody recognition site, useful for *in situ* hybridization or for Western blots,

(f) Epitope allowing a very stringent purification of the fusion protein using affinity columns. This sequence can also be at the C-ter end,

(g) Sequence containing the STOP codon for the fusion protein (which supposes that there is no other upstream STOP: The d, a, e, f and g sequences constitutes an open reading frame starting from the initiation codon of the sequence a up to the STOP codon in the g sequence)

(h) Terminator sequence specific of the cell type in which the plasmid is introduced (for example, in mammalian cells, the sequence BGHpA of the BGH gene / Bovine Growth Hormone / leading to a mRNA that is efficiently poly-adenylated and therefore very stable.

(i) selection genes allowing a positive selection screening of the transformed cells, these can be several genes allowing the transfer of the plasmid from bacteria (construction step) to the cells in which the analysis should be performed if these are not bacteria

(j) Replication origin, specific for the cell type in which the plasmid is introduced; there is also an origin in bacteria if the plasmid should be used in E. coli for the construction and multiplication steps.

Figure 1.11   Plasmid constructs containing specific sequences

primers are designed to fulfill this condition. One example of the type of plasmid designed to clone PCR product is given by the topo-plasmids® from Invitrogen. These vectors are linearized and the two 3′ extremities carry an overhanging T associated with a topoisomerase; this enzyme will efficiently favour the insertion and ligation of the PCR products themselves carrying at their 3′ ends an overhanging A added by the Taq pol at the end of the PCR elongation steps. The mix between the PCR products and the topo-plasmids leads in a few minutes to the exclusive insertion of the amplified fragment of interest with an efficiency reaching almost 95 per cent of the plasmids.

Box 1.8   Plasmids are efficient tools for cloning and analysing DNA sequences

Plasmids come from bacteria but now make up complicated constructions making efficient tools with multiple functions which are usable in many cell types. They have a cloning site where the sequence of interest can be inserted, one of several selection genes allowing the positive screening of the cells transformed by the penetration of this plasmid, and a eukaryotic or prokaryotic replication origin, allowing its autonomous replication in one or the other cell types.

Depending on the other sequences located on both sides of the cloning site, each type of plasmid allows a precise study of the target sequence, such as its sequencing, transcription or translation, *in vivo* or *in vitro*, to perform a functional or a structural analysis of the sequence product, to localize the product in the cell or, finally, to produce the product to allow it to be extracted and purified.

In the second case, the cDNA molecules are more heterogeneous than PCR products and two cDNA molecules obtained from the same mRNA can vary at their extremities depending on where their synthesis stopped. To obtain recombinant plasmids likely to incorporate a cDNA in frame with the other sequences of the fusion protein, a mix of three plasmids for which the sequence containing the insertion site for the cDNA is shifted by 0, 1 or 2 base pairs relative to the AUG initiation sequence, in a way that each cDNA will be in frame with the upstream sequences in at least one of the three recombinant plasmids. However, this means that all the sequences leading to a fusion protein have to be upstream of the cloning site, the STOP codon being the cDNA natural codon.

1.7   DNA fragment sequencing

The two main *in vitro* DNA sequencing methods were published in 1977, a few months apart, by Maxam and Gilbert and by Sanger *et al.* and will revolutionize in the same way as for PCR molecular biology and its applications in fundamental research as well as in diagnosis. Because it is simple and adaptable to the technological evolutions, the Sanger method is now universally used.

1.7.1   *Principle of the Sanger method: the sequencing reaction*

The Sanger sequencing method is applicable to a target fragment purified or specifically accessible. It is based on the possibility of performing *in vitro* replication of one of the two strands of the target, which supposes that a primer capable of hybridizing to the other strand that will be used as a template can be designed and added together with the trinucleotides and the DNA pol.

If the target fragment was obtained by PCR, its extremities are known and the problem of the sequencing primers is solved. If the target fragment is unknown, it is necessary to clone it into a vector whose sequences at the border of the cloning site are known and can be used to design primers.

The principle of the Sanger method consists of starting the synthesis of one of the two strands of the target fragment *in vitro* in the presence of small quantities of dideoxynucleotides (deoxy in 2′ but also in 3′) that will stop the elongation in the event that they are incorporated because in the absence of a 3′ OH, it is impossible to add another nucleotide (Figure 1.12). The analysis of the size of the different randomly interrupted synthesized fragments with respect to dideoxynucleotide incorporation (Figure 1.13) will allow the deduction of the sequence of the studied strand (the one synthesized starting from the primer). It is possible, and even recommended, to realize the sequence of the other strand in four other tubes with a primer b/− that can hybridize with the (+) strand that will become the template.

1.7.2 Reading of the sequencing reaction products

After the sequencing reaction, the goal is to measure, with single-base precision, the length of the newly synthesized strands, which can only be done by separating the fragments by high-resolution gel electrophoresis (precise separation of strands differing by one single nucleotide) and considering that the newly synthesized fragments can be distinguished from the other DNA fragments that are not relevant (primers, fragment of interest, etc.). Therefore, it is necessary for the newly synthesized strands to be labelled in a way so that they can be specifically identified.

There are two ways of labelling synthesized strands: using a labelled primer (which is done with universal primers corresponding to the sequences in the cloning vector), or using labelled dideoxynucleotides (see below for the labelling techniques). Even if radioactive labelling is now obsolete, the associated reading technique will be described in order to understand better the improvements made by fluorescent labelling and the use of automated sequencers.

Reading of a sequence with the labelled fragment emitting only one type of signal (radioactivity or cold labelling)

Using high-resolution polyacrylamide gel electrophoresis capable of separating fragments differing only by one nucleotide, it is possible to load on the gel the products of the four sequencing reactions and migrate them in four parallel lanes.

The DNA fragments will migrate according to their size and the revelation of the signal of the labelled molecules, meaning the newly synthesized strands whose size

| 5′ | Strand (+) | TCGATTGGCAGCATGGACGC | 3′ |

| 3′ | Strand (−) | AGCTAACCGTCGTACCTGCG | 5′ |

| Amplification primer a (+) /30 b |

TUBE 1: in the presence of ddT

| 5′ | + strand potential | TCGATTGGCAGCATGGACGC | 3′ |

Primer a (+) hybridized to the (−) strand
amplified: 30 nucleotides +?

Result: size of the neo-strands depending on the possible incorporation of a ddT:
30 + 1 or 5 or 6 or 14

TUBE 2: in the presence of ddC

| 5′ | + strand potential | TCGATTGGCAGCATGGACGC | 3′ |

Primer a (+) hybridized to the (−) strand
amplified: 30 nucleotides +?

Result: size of the neo-strands depending on the possible incorporation of a ddC:
30 + 2 or 9 or 12 or 18 or 20

TUBE 3: in the presence of ddG

| 5′ | + strand potential | TCGATTGGCAGCATGGACGC | 3′ |

Primer a (+) hybridized to the (−) strand
amplified: 30 nucleotides +?

Result: size of the neo-strands depending on the possible incorporation of a ddG:
30 + 3 or 7 or 8 or 11 or 15 or 16 or 19

TUBE 4: in the presence of ddA

| 5′ | + strand potential | TCGATTGGCAGCATGGACGC | 3′ |

Primer a (+) hybridized to the (−) strand
amplified: 30 nucleotides +?

Result: size of the neo-strands depending on the possible incorporation of a ddA:
30 + 4 or 10 or 13 or 17

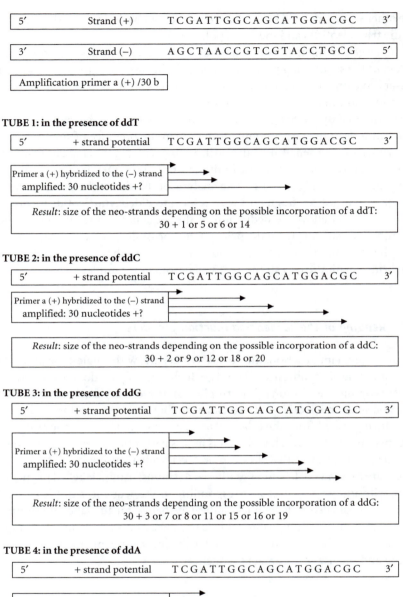

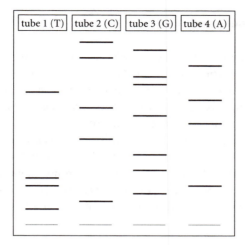

Reading of the gel autoradiogram, with the sequence of the (+) strand after electrophoresis of the products of the Sanger reaction, labelled with the same type of signal.
● The smallest fragment corresponds to the non-elongated primer (dotted line) that is only visible if labelled primers are used instead of labelled ddNTPs.
● A fragment one base longer appears in the lane corresponding to the first tube and indicates that the first base following the primer is a T.
● Then, moving up on the gel, it appears that the second base after the primer is a C (tube 2), then a G (tube 3), etc., giving TCGATTGGCAGCATGGACGC.

Figure 1.13   Reading of the autoradiogram

is the direct consequence of the sequence of the (+) strand, will allow the direct deduction of this sequence (Figure 1.13)

The labelling signal never allows direct reading of the gel; in the case of a radioactive labelling (low intensity) the gel should be applied on a negative that will be revealed a few days later; in the case of a cold labelling, the gel should be revealed and exposed with an autoradiogram (Figure 1.13).

● The DNA to be sequenced is dispatched in four tubes after being denatured (heated then the tubes are transferred to ice to prevent renaturation), together with the four tri-nucleotides, the DNA pol and the ad hoc buffer.
● In each tube, a small quantity of one of the four dideoxynucleotides, ddT (tube 1), ddC (tube 2), ddG (tube 3) and ddA, (tube 4)
● In each of the four tubes, the primer of 30 nucleotides will hybridize with the (−) strand and will be elongated to form eventually a (+) strand whose sequence is as indicated above. However, this elongation will be stopped as soon as a ddN is incorporated.
   Due to the sequence of the (+) strand to synthesize, indicated on top of the primer, the size of the fragments will never be the same in the four tubes, as it is clearly shown in the result relative to each tube.
● To deduce the sequence, it is sufficient to 'know how to measure' the size of the different fragments synthesized in each of the four tubes (Figure 1.13).

Figure 1.12   Sanger sequencing reaction

Box 1.9   DNA sequencing

The most used sequencing method is the Sanger method. It consists of randomly interrupting the elongation of a primer specific for the target sequence by incorporation of a dideoxynucleotide. Then, using high-resolution electrophoresis as a separation system, it is possible to separate the fragments obtained by addition of one, two or *n* nucleotides before the interruption by incorporation of the dideoxynucleotides which form the extremities of the elongated primers.

Because the labelling differs depending on the incorporated dideoxynucleotide (depending on whether it carries a cytosine, thymine, adenine or guanine), the visualization of the fragments separated by the electrophoresis is able to determine the incorporated nucleotide, which corresponds to the sequence on the DNA strand.

Reading of a sequence with a fluorescent labelling and use of an automated sequencer

The advantage of fluorescent labelling relies on the fact that different signals can be emitted as opposed to radioactive labelling which only emits one type of signal. If the four ddNTPs are labelled with a different fluorochrome, the Sanger reaction can be performed in a single tube (saving time and reagents) because the different types of newly synthesized strands will be labelled in a different way depending on whether they are terminated by a ddT, a ddC, a ddG or a ddA. Therefore, the electrophoresis can be done in a single lane (Figure 1.14), the fragments being separated by migration on the gel. The latter is fixed in a vertical orientation in an apparatus containing a laser which keeps sending an incident ray in an open window at the base of the gel in a way that any labelled fragment going through the window will have its fluorochrome excited and will emit, at its specific wavelength, a signal that will be captured by a receptor. This receptor is coupled to a computer which will reconstitute the sequence obtained by the succession of fluorochrome types emitted by the different fragments when they pass in front of the excitation and reception window. The computer is also set up to give the decoded sequence, together with the intensity of the emitted fluorescence of each strand passing through the laser window (Figure 1.14)

Because the sequencing products do not need to migrate in parallel lanes, it is not necessary to perform the electrophoresis in flat gels, and instead a gel confined in a capillary tube can be used, which is really adapted to diagnosis by sequencing when the number of sequences is small and variable (the washing of the capillary, pouring of the gel, loading of the sequencing products, laser emission, signal reception and analysis are automated: it is sufficient to put a few microlitres of the sequencing reaction

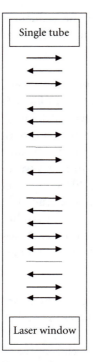

- In a single tube, the fragments ending in a different ddN carry a different fluorochrome. The four fragment types are shown schematically here by:

  ↔ if ddT, → if ddC, ← if ddG, — if ddA

- They migrate together on the same lane with a speed corresponding to their respective length.
- They pass successively in front of the laser window, the first one (the shorter) carries the fluorochrome specific for ddT, then the one carrying the fluorochrome specific for the ddC passes, etc., giving: TCGATTG-GCAGCATGGACGC.

← Reading, by an automated sequencer, of the sequence of the (+) strand after electrophoresis of the products of the Sanger reaction, labelled with a different fluorochrome on each ddNTP

Figure 1.14   Reading with an automated sequencer

into the apparatus and to push the button to get the sequence out about 30 minutes later).

Note 1 If the primers are labelled with a single nucleotide, it is necessary to perform four reactions and then four migrations in parallel because the fragments are labelled by the same signal, this one being read at the bottom of the gel when they pass in front of a window covering the whole width of the gel. If a set of primers labelled with four different fluorochromes is available, it is still necessary to perform the reaction in four different tubes in order to keep separate the newly synthesized strands ending in a ddT, a ddC, a ddG or a ddA, but the contents of the tubes can be pooled for the reading of the sequence on the gel.

Note 2 The sequencing of a DNA fragment is completed when the sequences of both strands have been performed and compared to verify their complementarity.

Note 3 The trace of the fluorescence emission is very useful, either to remove some uncertainties about the complementarity of the two strands, or to identify point mutations, for example in a heterozygote where the two allelic fragments of a gene have been co-amplified by PCR and then co-sequenced. In such cases, at the level of the point mutation, two newly synthesized strands with the same size will pass through the window and emit a different wavelength. The computer coupled to the automated sequencer is able to interpret quantitatively emissions of equal intensity, for example 'green + red', and will indicate the sequence 'A + T', meaning that one strand carries A and the other T; in other cases, it cannot do so, and will indicate in this case the letter 'N' for nucleotide; the experimenter will have to take over to interpret and remove the ambiguities seen by the sequencer.

Automated sequencers with flat gels are also useful because their dimensions adapt them to the sequencing of large series of sequences (about 30 lanes in parallel, corresponding to 30 sequences).

It should also be noted that automated sequencers are extremely well adapted to measure the size of PCR fragments, which makes them very effective in all diagnoses based on such measures (repeat sequences in triplet disorders or genetic imprinting, ARMS and OLA techniques, see below).

The automation of the sequencing associated with the automation of the sequencing reaction (more difficult) has made it possible to save a lot of time, which is necessary in order to perform genome sequencing.

1.8 Modification of the sequence of a DNA fragment: site-directed mutagenesis

In vitro site-directed mutagenesis of a gene is the first step towards directed mutagenesis of an organism by replacing one of the two copies of a gene by a gene copy mutated *in vitro*.

A simple method among others consists of the cloning of the sequence of the wild type cDNA of the gene of interest into a suitable plasmid. A purified preparation of this plasmid is then submitted to a PCR using parallel primers carrying a single-base difference where the sequence should be modified, compared with the wild type cDNA sequence. The PCR conditions are defined to tolerate a mismatch between the 'mutated' primers and the wild type sequence during the hybridization step, at least during one or two cycles; the hybridization can become more stringent once a few copies of the plasmid carrying the mutated sequence have appeared.

This way, all the amplified plasmids are carriers of the mutated sequence; the plasmids are individually cloned and re-isolated from a single clone and sequenced in order to test if the chosen clone carries the mutated sequence of the gene.

1.9 Molecular hybridization techniques and applications

1.9.1 Introduction

The principle of molecular hybridization techniques relies on the ability of complementary single-strand DNA (or RNA) to renature spontaneously to form a double-stranded molecule with lower free energy and therefore thermodynamically favoured.

This spontaneous annealing depends of course on the level of complementarity; two sequences that are strictly complementary will renature more easily than two sequences with a local variation. The local variation will introduce one or several mismatches in the duplex which will destabilize it by lowering the denaturation temperature.

Depending on the level of complementarity, the renaturing and the stability of the duplex will then depend on the temperature which will be higher the more complementary the sequences to anneal (hybridize) are.

The PCR is a technique based on molecular hybridization because priming is the result of annealing the primers with the extremities of the fragment to amplify. The primers are made of synthesized oligonucleotides whose sequences are strictly complementary (except in special cases, see site-directed mutagenesis or ARMS). The primers' sequences are determined in such a way that the hybridization temperatures are as close and as high as possible, precisely in order to limit 'illegitimate' renaturing with other genomic fragments that would lead to a parasitic amplification.

The hybridization techniques are also used to identify a particular fragment within a mix of fragments, and to study its size and possibly mutations of the sequence. In this case, a known and purified DNA fragment is used as a probe. The probe is labelled, meaning it is capable of emitting a signal, and the fragments are incubated with the probe in order to reveal a possible hybridization with the target fragment that might be there and that would therefore be indirectly labelled.

Of course, the fragments from the mix to be studied would have to be denatured, as well as the probe if it is double stranded, in order to allow its hybridization with the target fragment if present in the mix.

1.9.2 Probes, labelling and reading of the signal

Depending on the circumstances, probes can be a fragment of genomic DNA from a DNA library cloned into a vector, or a cDNA also isolated from a clone from a cDNA library. However, a synthesized oligo probe, or an RNA obtained by *in vitro* transcription of a DNA fragment cloned in a plasmid carrying a promoter next to the cloning site and allowing this manipulation, can also be used.

The labelling techniques depend on whether the probe is single or double stranded. The radioactive labelling of a probe allows the emission of a signal allowing detection of the fragment recognized by hybridization by autoradiography on any substrate. Non-radioactive labelling (cold labelling) of a probe does not make it capable of emitting a signal but simply gives it the capability of binding to a system emitting the signal.

In order to detect a fragment recognized by the labelled probe on any substrate, it is necessary to insert a reading step that will add the emission system to the probe, in order to reveal its hybridization by the presence and the measurement of the emitted signal. This signal can be visualized directly (pigment visible by the eye) or by autoradiography (photon emission).

Labelling of a synthesized oligoprobe

A synthesized oligoprobe is single stranded – due to the way it has been synthesized it carries a 5′OH extremity instead of a 5′P. Radioactive labelling can take advantage of this characteristic to use a polynucleotide kinase together with ATPγ^{32}P, allowing phosphorylation of the probe.

Radioactive labelling can also be performed using a terminal transferase in the presence of ATPα^{32}P, allowing the formation of a radioactive polyA tail at the 3′ end of the probe. This kind of labelling is used when a very intense signal is required.

The second type of labelling is specific for cold labelling, where the terminal transferase acts in the presence of UTP with the uracil carrying a molecule (biotin or digoxygenin being the most common) on to which the signal emission system can be bound (Figure 1.15).

Labelling of a genomic probe or a cDNA probe

A double-stranded genomic or cDNA probe can be labelled using one of three methodological principles.

1. The substitution of a nucleotide from the probe with a labelled nucleotide, radioactive (for example ATPα^{32}P) or cold (UTP with the uracil carrying the biotin of digoxygenin). This technique (Nick-translation) consists of making small cuts with DNAse I on the double-stranded DNA and, after inactivation, using all generated 3′ extremities to add the DNA polymerase I together with the four dNTS and the ATPα^{32}P that will allow the incorporation of radioactive adenosine (or biotinylated uridin) during the reparation of the duplex.

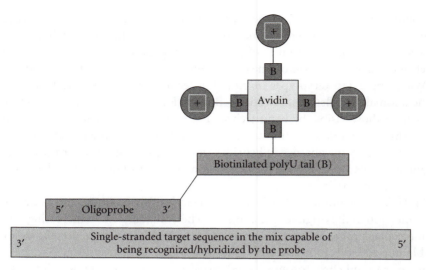

- The probe that is hybridized to the fragment of interest is labelled by the addition of a biotinilated polyU tail (the digoxigenin can replace the biotin); the schema only shows one molecule of biotin from the tail.
- The presence of the probe, hybridized to the fragment of interest is revealed in several steps:
  (a) addition of avidin, molecule carrying four sites of fixation for the biotin (for the digoxigenin labelling, an anti-digoxigenin is used); each molecule of biotin binds a molecule of avidin;
  (b) addition of free biotin, carrying a molecule able to emit a signal (+, fluorescence for example) or an enzyme like alkaline phosphatase;
  (c) the emitted signal is measured by measuring the fluorescence if the free biotin carries a fluorochrome, by autoradiography after adding a substrate for the alkaline phosphatase capable of emitting photons if the free biotin (anti-digoxigenin antibodies) is carrying this enzyme.

Note: It is possible to amplify a signal using the 'sandwich' technique that consists of the superimposition of several avidin–biotin free layers before adding the biotin carrying the signal.

Figure 1.15   Cold labelling and revelation principles

2. *In vitro* synthesis where copies of the probe are made in the presence of labelled nucleotides. This technique (random priming) consists of denaturing the probe to generate two single-stranded fragments that will be used as templates for DNA polymerase I to synthesize labelled copies of these fragments in the presence of random hexanucleotides that can renature locally with each fragment to serve as primers, and of the four dNTP and ATPα^{32}P (or UTP carrying biotin or digoxygenin). This second technique is the more efficient of the two.

3. The lengthening at the 3′ end by a terminal transferase to form a labelled tail (see above).

1.9.3   *FISH* and in situ *PCR*

Fluorescent *in situ* hybridization (FISH) is a molecular hybridization technique with one or several probes each labelled with a different fluorochrome. These probes can be antibodies or DNA probes and their targets can be cellular or sub-cellular structures that can be observed by fluorescence microscopy. FISH allows an analysis at a scale that lies between that of the nucleotide sequence or peptide sequence, at the molecular level, and that of the chromosome or the cellular fraction, at the microscopic level.

When applied to karyotypes (*in situ*), this technique allows the study of chromosome anomalies (deletions, insertions, duplications, translocations) that are much finer than those visible by band techniques, even at high resolution (see Chapter 4). In this way, a microdeletion undetectable at the level of the R and G bands can be discovered by the absence of hybridization of a probe specific for the concerned region associated with a fluorescent emission of a control probe hybridized at a site next to the microdeletion site.

FISH does not only allow a reduction of the scale of analysis but it can also allow the identification of the origin of the anomaly. The availability of an increasing number of molecular probes specific for a particular chromosome sequence allows a non-ambiguous identification of the chromosome origin of the translocated fragment or of the extra mini chromosome (called 'marker' in cytogenetics), which is essential for the diagnosis and prognostic evaluation.

The *in situ* karyotype analysis can be performed at an even smaller scale by *in situ* PCR which allows, after partial DNA decondensation, the amplification of a fragment of interest in the genome, before the hybridization of a labelled oligoprobe specific for the amplified fragment is tested. However, *in situ* PCR is more often used together with FISH to study tissues or slices, notably to detect mRNA or viral RNA.

1.9.4   *Detection and dosage methods using signal amplification*

These are detection and dosage methods which, instead of being based on the amplification of the target sequence, rely on the amplification of a signal whose emission

---

Box 1.10   Fluorescence *in situ* hybridization (FISH)

Molecular hybridization between complementary DNA sequences has been used to develop a new technology allowing the visualization of biological material at a scale that lies between the microscopic and the molecular scale, inaccessible to microscopic visualization. Each cellular or sub-cellular structure visible with the microscope and containing a nucleic acid molecule (chromosome, nucleus or ribosome) can be treated *in situ* after fixation on a slide that makes the nucleic acid molecule (RNA or DNA) accessible for a specific fluorochrome-carrying probe. The fixation of the fluorescence shows that the specific sequence is present and allows its localization. The absence of fluorescence shows that the sequence is absent. FISH brings an important complementary technique to the cytogenetic or genetic analysis because it gives information on the chromosome localization of a sequence, whether it has been deleted or translocated.

depends on the presence, even in minimal quantities, of the target sequence in the biological test sample (for example, detection of the HIV virus or the hepatitis B or C virus in serum).

The principle of these methods (bDNA which means branched DNA) consists of the capture of the DNA on a solid surface by partial hybridization with probes fixed on that surface. In a second stage, a first wave of probes is fixed by hybridization to the possible target sequence where they are not hybridized to the capture probes, then a second wave is fixed to the first wave probes so that an increasing amount of DNA probes 'branched' one to the other accumulates, starting from the target sequence, with the last ones being labelled. To a unique sequence constituting the base of the 'branched DNA tree' are associated a large number of signals emitted by all the extremities of its multiple arborescences.

1.9.5   Southern blot hybridization

A Southern blot is made up of a collection of restriction fragments that have been separated by electrophoresis, denatured and transferred onto a solid membrane where they are accessible to molecular hybridization techniques (see Figure 1.5).

The transfer from the gel onto the membrane does not modify the electrophoresis migration distances, allowing the measurement of the size of the fragments depending on the migration pattern of the size marker that co-migrated in the gel.

The technique of molecular hybridization is used for the study of a *Southern blot* to identify among thousands of restriction fragments, the fragment(s) of interest for the study or diagnosis that are undistinguishable from the others in the smear visible on the gel (Figure 1.4 and photo).

If a probe sharing a common sequence with the fragment(s) is available, it is possible to hybridize the membrane (the fragments are single stranded) with the labelled probe and then to detect the signal of hybridization of the probe to localize the target fragment(s) (Figure 1.16). A use for this method is shown by the diagnosis of fragile X syndrome (Figure 2.13) or haemophilia (Figure 2.16).

It is worth noting that after the annealing stage, the membrane should be washed in order to leave only the probe molecules and leave the membrane free of any non-specific adsorption of the probe. This washing step is always critical with hybridization techniques; it determines the quality, sensitivity and reliability of the analysis. The 'stringency' determines the strength with which the washing of the membrane is performed. The stringency combines a set of three parameters that affect the 'release' of the probe: the temperature, the ionic strength and the denaturing agents. The stringency of the washes increases with increases in temperature (thermal effect on the hydrogen bonds of the duplex or on the adsorption forces), with decreases in ionic strength and with increases in detergent concentration. By acting on these three parameters, the good washing conditions for a given Southern blot can be empirically determined. The same factors apply for a Northern blot.

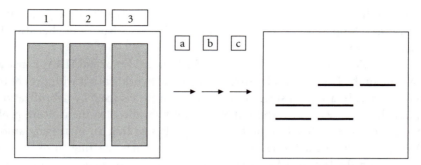

- On the left: observation, using UV transillumination, of the electrophoresis gel showing the smear left by the migration of the restriction fragments obtained by digestion of the genomic DNA of three individuals.
- The three arrows correspond to the three successive treatments applied to these fragments (Figure 1.5):
  (a) denaturing and transfer onto the membrane (Southern blot);
  (b) hybridization of the Southern blot with a probe containing a fragment of genomic DNA;
  (c) washing of the Southern blot and revelation of the specific labelling by annealing of the probe to the target fragment identified in the three smears.
- On the right: observation of the target fragments 'recognized' (hybridized) by the labelled probe, among thousands of fragments on each smear.
  Analysis: considering the pattern of the gel realized by co-migration of the ladder, it is possible to measure the size of the recognized fragments:
  - from the first individual: two fragments of 3 and 2 kb,
  - from the second individual: three fragments of 5, 3 and 2 kb,
  - from the third individual: a fragment of 5 kb.
  This is the picture generally observed for three individuals with a different genotype concerning the presence or the absence of a restriction site for the enzyme used to perform the genomic DNA digestion, in the region of the genome specifically recognized by the probe (see below in section 1.9.8, RFLP)

Figure 1.16   Identification of target fragments on a Southern blot after hybridization by the probe

1.9.6   ASO techniques: dot blot and reverse-dot blot

The study of DNA fragments amplified by PCR does not raise the same problems as with the study of restriction fragments on a Southern blot where the goal is to identify one (or several) target fragments in a smear and to measure its (their) size.

Indeed, PCR gives an almost pure solution of the target fragment(s) so, therefore, its (their) analysis is easier:

- they are easily separable by electrophoresis and their size is directly measurable on the control gel;

- they can possibly be sequenced immediately after the PCR;

- they can also be studied for possible variations in their nucleotide sequence (SNP stands for single nucleotide polymorphism), notably when the PCR, being performed using diploid tissue to start with, allows the co-amplification of the target fragment from the two homologous chromosomes (allelic variation).

Box 1.11   DNA blotting

The identification of a target sequence can require hybridization with a specific probe that will reconstitute a double-strand molecule. The Southern blot consists of transferring DNA fragments separated by electrophoresis and denatured onto a nylon membrane; in this case, the target fragment(s) is(are) recognized by a specific labelled probe. The transfer is called a Northern blot when the molecules transferred after electrophoresis are RNA and it is called a Western blot when it concerns peptides.

A dot (or slot) blot is made up of a nylon membrane on which is deposited a certain amount of target sequence prepared by PCR, so it is chemically homogenous (no need for electrophoresis in this case); this dot or slot blot can then be hybridized with a labelled oligoprobe in order to test the conformity of the target sequence with the oligo probe (allele specific oligonucleotide).

The reverse dot is a membrane on which a set of oligoprobes are bound and that is hybridized with the PCR products obtained from DNA. This allows testing in one single assay, the presence or absence of point polymorphisms in the PCR products depending on whether or not they can hybridize with the oligoprobes.

An allelic variation (for example the mutation in a gene) can be easily and directly detected if it is a size modification (deletion or insertion of one or several nucleotides, see ΔF508 mutation in the CFTR gene; amplification of a repeat sequence, see fragile X syndrome), measuring the size of the amplicons is sufficient.

An allelic variation can also be easily detectable if it creates or removes a restriction site (see haemochromatosis or haemoglobinopathies); it is sufficient to test the size of the PCR fragments after digesting them with the relevant enzyme.

If the allelic variation of the nucleotide sequence in a precise site of the amplified DNA fragment does not present any characteristic making it directly observable on a control gel (as described above), molecular biologists use, in some cases, allele specific oligonucleotides (ASO) oligoprobes in order to test their possible hybridization with the PCR fragments loaded on a membrane (dot blot or slot blot after having been denatured (see Section 1.5.1 above).

The ASO probes, single stranded, with an average size of 20 nucleotides, are determined with respect to the standard known sequence of a gene segment and to the sequence or the different sequences of the mutated segment of this gene (which supposes that these mutations have been previously identified). They allow the detection of the presence of a mutation on the amplified and tested segment (Figure 1.17) depending whether or not they hybridize with the PCR products.

This 'all or nothing' answer (hybridization or non-hybridization of the ASO to the PCR products) is possible because the stringency defined for the washing of the dot

- Each 'spot' on the dot blot corresponds to the PCR products of exon 1 of a gene involved in a recessive disease, performed on the healthy parents healthy carriers (father: spot 1, mother: spot 2), their affected child (spot 3) and the child to be born (fetal DNA from trophoblast cells: spot 4).
- On the left, display of the signal on a dot blot hybridized with a normal ASO probe for a 20 base segment of the tested exon 1.
- On the right, display of the signal on a dot blot hybridized with an ASO probe specific for a known mutation m1 on the 20 base segment of exon 1.

Conclusions

- The parents 1 and 2 are both carriers of the same mutation m1. Because we know they are healthy carriers, their genotype is N/m1 (where N is the 'normal' or functional sequence of the studied gene).
- *Note*: the presence of a normal allele in the parents is deduced from their healthy phenotype and not from the normal ASO hybridization that tests 20 bp of the gene and not the entire gene (see Figure 1.18).
- The affected child is homozygous m1/m1; none of its two genes have the normal 20 bp segment so there is no hybridization with the normal ASO.
- The expected child carries at least one mutated gene (hybridization with the mutated ASO). The hybridization with a normal ASO shows that the other copy of the gene does not carry the m1 mutation, and considering its parents genetic situation, it proves that this copy is obligatory N. The expected child is a healthy carrier.

Figure 1.17   Analysis of a dot blot using ASO

blot allows only the ASO probe showing a perfect complementarity with the PCR product tested to be kept.

However, it is important to understand the limits of the use of ASO probes, and consequently the limits of the efficiency of the diagnosis associated with their usage (Figure 1.18):

- because they are synthetic probes, the ASO probes only allow testing of gene mutations that are already known;

- the absence of hybridization of an ASO probe specific for a mutation signifies that the gene fragment, amplified by PCR and tested with the ASO probe, does not carry the mutation but it does not mean that the gene is not mutated; the ASO probe only tests 20 nucleotides in the gene sequence – the gene can carry a mutation elsewhere, on a sequence not covered by the probe.

The ASO technique is easy, reliable, inexpensive, and fast when there is only one gene or one sequence that is not too polymorphic (one or a few mutations), but can get expensive (a dot blot by probe, a hybridization, washes and a revelation) and less efficient when the number of allelic variations becomes large because it is necessary to design a pair of 'standard' (or normal) and variant (or mutated) ASO probes for each allelic variation.

- Each 'spot' on the dot blot corresponds to the PCR products of exon 1 of a gene involved in a recessive disease, performed on the parents (carriers: father, spot 1, mother: spot 2), their affected child (spot 3) and the child to be born (fetal DNA from trophoblast cells: spot 4).
- On the left, display of the signal on a dot blot hybridized with a normal probe ASO for a 20 base segment of the tested exon 1.
- On the right, display of the signal on a dot blot hybridized with an ASO probe specific for a known mutation m1 on a 20 base segment exon 1.

Conclusions

- The parent 1 carries the mutation m1. Because we know he is a healthy carrier, his genotype is N/m1 (where N is the 'normal' or functional sequence of the studied gene).
  Note. The presence of a normal allele in the parents is deduced from their healthy phenotype and not from the normal ASO hybridization that tests 20 bp of the gene and not the entire gene (see Figure 1.17).
- The parent 2 is not a carrier for the mutation m1, which can be seen by the absence of hybridization with the mutated ASO, she is therefore a carrier for another mutation X, detectable using another pair of ASO probes if it is known, or not detectable if it has not been identified yet. The genotype of this parent is N/X.
- The affected child is a compound heterozygote m1/X (each copy of the gene carries a different mutation); the hybridization with the normal ASO only signifies that the gene carrying the X mutation has a normal nucleotide sequence in the 20 bp segment tested by this ASO.
- The expected child carries one mutated copy m1 (hybridization with the mutated ASO) transmitted by the parent 1. The hybridization with a normal ASO shows that the other copy of the gene does not carry the m1 mutation, and considering the genetic background, it is impossible to reach a conclusion (the same way it has been done in figure 1.17). Indeed, whether the allele transmitted by the parent 2 is normal (N) or mutated (X), the normal ASO will hybridize anyway because N and the gene carrying the mutation X have a normal nucleotide sequence in the 20 bp segment tested by this ASO.
  The expected child has a one in two risk of being affected and a one in two risk of being a healthy carrier.

Figure 1.18   Limits in the analysis of a dot blot by ASO

This is the reason why the *reverse-dot* techniques have been developed, techniques in which several ASO probes specific for several different mutations of a gene are bound to a membrane and then hybridized with the PCR products used as mobile probes after they have been labelled.

In addition, the development of multiplex PCR allowed the co-amplification, in a single tube, of several different fragments of the same gene, bringing simultaneously pairs of primers specific for the fragments to be amplified. Multiplex PCR primers must be designed in such a way that their annealing temperatures, which are sequence dependent, are as close as possible; however, it is possible to add to the PCR mix a molecule that attenuates the effect of the sequence on the hybridization temperature.

The co-amplification of many segments of a gene using multiplex PCR, their labelling and their use as mobile probes on a single reverse dot membrane, leads to a considerable reduction in the cost (both in time and in consumables) and allows the testing in one single experiment of 10 to 12 different mutations in a gene simultaneously.

1.9.7   ARMS and OLA techniques

The reverse-dot technique has been developed to test simultaneously for the presence or the absence of a large number of polymorphisms (HLA typing) or of potential mutations of a gene. However, the evolution of the technologies has made it possible to save even more time and consumables by skipping the steps of deposit onto the membrane, hybridization and revelation and by replacing them by a direct visualization of the multiplex PCR amplification products on an electrophoresis gel.

Amplification refractory mutation system (ARMS)

The principle of this method consists of the amplification of a short fragment of a gene using on one side a 'common' primer and, on the other side, two primers that have at their 3′ terminus a base complementary to that present on the standard sequence of the gene (standard primer), or for the base present on the mutated sequence (mutated primer) (Figure 1.18). Also, these two primers are elongated so that their length should be different.

For an individual whose two genes are not carrying the tested mutation, the PCR will function with the common primer and the standard primer and will give amplimers with a fixed length visible on a gel (Figure 1.19).

For an individual carrying two identical mutated copies, the PCR will only function with the common and mutated primers, giving amplimers of different lengths on the gel (the mutated primer being longer or shorter than the standard primer).

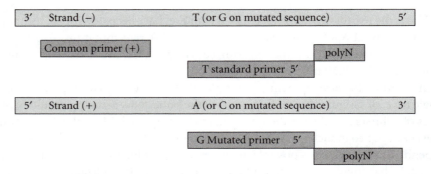

- A common primer amplifies the (+) strand (it hybridizes with the (−) strand).
- The standard primer amplifies the (−) strand if it is not mutated and the mutated primer amplifies the (−) strand if it is mutated; each of the two primers carrying different sizes polyN or polyN′ tails.
- The DNA fragment amplified by PCR will be longer if the tested DNA is mutated and shorter if it is not, the size difference being visible on an electrophoresis gel.

Figure 1.19   Principle of the ARMS

On an individual heterozygous for the mutation, the three primers will take part of the PCR and the gel will allow the visualization of the two amplified fragments with different size.

The possibility to realize multiplex PCR makes it possible to test simultaneously several mutations by setting up the PCR with a triplet of primers (common, standard and mutated) per mutation to test, making sure that the length of the primers are quite distinct for the analysis of amplimer size not to be ambiguous.

The oligonucleotide ligation assay (OLA) technique

This more recent technique is even more effective than ARMS because it can test a larger number of mutations simultaneously thanks to an automatic reading of the length of the fragment obtained when it passes by the reading window of an automated sequencer.

In the first step, a multiplex PCR is performed for several segments of the gene on which the presence of a mutation should be tested (for example, 15 segments in the CFTR gene involved in cystic fibrosis). During the second step, several ligation cycles are performed using a thermostable ligase and primers allowing testing for the presence or absence of mutations.

The principle for the ligation test is as follows. For any point mutation tested, a common primer (carrying a fluorochrome) is designed for one side of the strand and two primers capable of hybridizing with that same strand adjacent to the common primer are designed for the other side. These two primers have a different nucleotide at their 5' end, one corresponding to the standard site, and the other to the mutated site; in addition, the two adjacent primers, standard and mutated, are respectively carrying a polyN tail and a polyN' tail of different sizes (Figure 1.20).

If the sequence of the tested strand is standard (not mutated), the standard primer will be able to hybridize normally while the mutated primer will not, and the ligase will be able to join the two primers to form a fragment of a defined size. If the strand is mutated, the common primer will be joined to the mutated primer generating a fragment of a different length.

A series of n denaturation/annealing/ligation cycles will allow the generation of a large number of ligation fragments. The products will be submitted to electrophoresis, in general in a capillary gel, and the fragments will migrate according to their size and pass a laser window that will provoke the emission from the fluorochrome; the co-migration of a size marker carrying a specific labelling allows the determination of the size of the fluorescent fragments passing by the laser window. The presence or absence of the tested mutation can then be deduced from the sizes seen (a ligation fragment whose size equals the sum of the sizes of the common and adjacent primers,

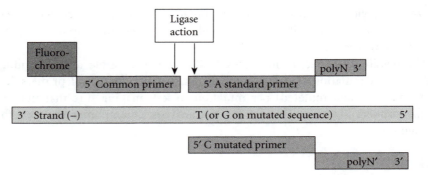

- The (−) strands from the PCR are hybridized with a common primer carrying a fluorochrome, just upstream of the tested mutation site, that carries a thymine (T) in the normal allele and a guanine (G) in the mutated allele.
- A primer immediately adjacent can also hybridize with the (−) strand, whether a standard primer with an A at its 5′ end and a polyN tail if the strand is not mutated, or a mutated primer with a C at its 5′ end and a polyN′ tail that have a different length if the strand is mutated.
- The ligase will, depending on the case, join the common primer with the adjacent primer, standard or mutated, generating a labelled fragment (fluorochrome from the common primer) whose size will depend on the polyN or polyN′ tail brought by the adjacent primer.
- Several denaturation/annealing/ligation cycles are performed (thermostable ligase).
- The labelled ligation products are analysed by electrophoresis (capillary or gel), in the presence of a size marker whose co-migration allows the deduction of the size of the fluorescent fragment passing in front of the laser window.
- The presence of a mutation is shown by the presence of a ligation fragment whose size equals the size of the common primer plus the size of the mutated primer.

Figure 1.20   Principle of OLA

standard or mutated). All the standard fragments and mutated fragments obviously have different sizes (see simultaneous test for the 31 mutations of the CFTR gene, Figure 2.3).

Box 1.12   ARMS or OLA are efficient tools to detect single nucleotide polymorphism (SNP)

A gene mutation or, more generally, a point polymorphism can be identified using other methods than hybridization with a specific probe. The amplification refractory mutation system (ARMS) method identifies a point mutation by performing a PCR using a set of three primers, one labelled and the other two that differ from each other by their length and most of all by their hybridization with the wild type sequence or with the mutated sequence; depending on the size of the PCR product, the presence or the absence of the mutation can be determined.

> **Box 1.13   Restriction fragment length polymorphisms are informative markers in genetic analyses**
>
> If two homologous DNA molecules differ by the presence or absence of a re-striction site at a particular position of the sequence, the digestion of these two molecules by the restriction enzyme will give two small fragments if the site is present and one large fragment if the site is absent, that is to say a restriction frag-ment length polymorphism. This polymorphism behaves like a di-allelic marker for its transmission ('+' presence of the site and '−' absence of the site) and any individual is a carrier, for this marker, of one of the three possible geno-types +/+, +/− and −/−. This type of marker is very useful in research and its applications.

1.9.8   Definition, analysis and applications of RFLPs

Definition

Among the large number of restriction sites that cover the genomic DNA, some are quasi-constant, meaning they are always present at the same locus on all homologous molecules from the species. In contrast, some sites are facultative or occasional and display molecular polymorphism both intra-individual (each individual being diploid and carrier of two homologous molecules for each chromosome pair) and inter-individual. In contrast to allelic polymorphism of the genes that concern the expressed part of the genome, the polymorphism of the absence/presence of the restriction sites covers the whole genome, in both the expressed as well as the non-expressed sequences.

Analysis methods of an RFLP site

It is possible to visualize the presence or the absence of a given restriction site at a particular genomic locus by their effects in the case of total digestion of genomic DNA by the restriction enzyme. Indeed (Figure 1.21(a)), the presence/absence of a restriction site leads to a polymorphism in the size of the restriction fragments (RFLP stands for restriction fragment length polymorphism) visible on a Southern blot (Figure 1.21(b)) if a genomic probe capable of identifying the fragments is avail-able, or on a PCR control gel, after digestion, if a genomic fragment containing the facultative restriction site can be amplified (Figure 1.21(c)). This allows the genotyp-ing of an individual by defining the genotype for the RFLP studied as homozygote −/− if the two homologous DNA molecules are not carrying the facultative site, or heterozygous +/− if it is present on one and absent on the other.

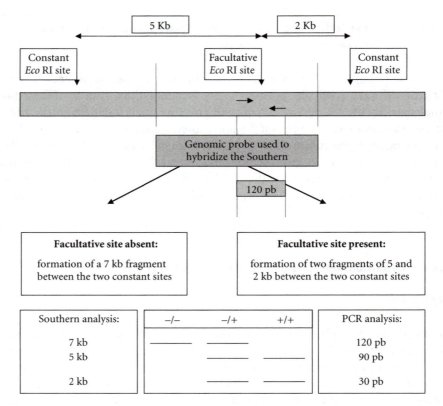

(a) A facultative site for *Eco*R1 is located between two constant sites at 5 and 2 kb respectively. Its presence or absence leads to a polymorphism in the length of the restriction sites in this region of the genome after complete digestion of the DNA.

(b) The size of these fragments can be seen on a Southern blot: after complete digestion of the genomic DNA, electrophoresis on an agarose gel, transfer onto a membrane and hybridization with a labelled probe that has partial homologies with the three types of fragments (see dotted lines on the top of the figure).

   The analysis of the number and size of the fragments allows the genotyping of the individuals studied, here from left to right, as −/−, +/− and +/+.

(c) If the genomic fragment containing the RFLP had been sequenced at least at its extremities, it is possible to amplify it by PCR (arrows corresponding to the primers in the figure under the facultative site).

   The PCR uses the complete digestion of the genomic DNA, the electrophoresis and the Southern. After the PCR, the amplified fragments are digested and visualized on a control gel (polyacrylamide or agarose). In the present example, a 120 bp fragment has been amplified with primers located 90 to 30 bp away from the facultative site, leading to a 120 bp fragment in case of absence of the site, and of 90 bp and 30 bp fragments in case of presence of the site.

Figure 1.21   Definition, consequences and vizualization of an RFLP

Application to the indirect genetic diagnosis of a monogenic disease

RFLPs constitute a very useful tool because of their efficiency and reliability for the indirect (or family) diagnosis of a monogenic hereditary disease, by analysis of the co-transmission of the mutations and the RFLP linked with the gene (Figure 1.22).

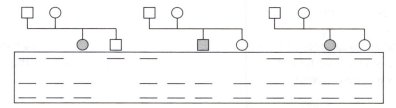

Genotypes of the individuals at the locus of the gene involved in an autosomal recessive disease, deduced from their clinical phenotype and, for the fetus, of the RFLP alleles inherited from the parents and co-transmitted with the normal allele (N) of the gene or with the mutated (m) pathogenic allele. (The presence or absence of the RFLP site at the vicinity of the gene is noted '−' and '+' respectively.)

(N,−) (N,−)(m,+) (N,−)   (N,−) (N,+) (m,+)   (m,+)   (N,−) (N,+) (m,+)   (?,−)

or

(m,+) (m,+)(m,+) (N,−)   (m,+) (m,+) (m,+)   (?,+)   (m,+) (m,−) (m,−)   (?,+)

- The parents are healthy carriers, their affected child appears in grey and the prenatal diagnosis is performed on the fetal DNA of the second child to be born.
- The gene involved cannot be directly studied if it has not already been cloned and sequenced, or if some of the parents carry unidentified mutations (see limits of the ASO, ARMS and OLA techniques), but an RFLP is known at the vicinity of the gene, with fragments that can be seen (the large fragment corresponding to the absence of the RFLP site, and the two short corresponding to its presence).
- The family study should start from the RFLP analysis of the affected child (if the DNA is not available, this strategy is not applicable); the child is necessarily mutated m/m on its two chromosomes and, therefore, each parental mutation can be associated with one of the two RFLP alleles observed in the proband, which makes it possible to deduce from the RFLP alleles transmitted to the studied fetus, which of the alleles N or m has been transmitted (this assumes that the risk of having a crossing over is minimum; otherwise, it has to be taken in account and the diagnosis result should be given with an error risk).
- *Family 1.* The affected child is m/m for the gene and +/+ for the RFLP, therefore, he is (m,+)/(m,+); the parents then have a chromosome (m,+) and the another chromosome that can only be (N,−) because they are healthy carriers, and they have a chromosome lacking the RFLP site. The fetus being −/−, must be N/N, homozygote for the normal allele, due to the co-transmission, and therefore will not be affected. *Note*: in each case of transmission of a parental RFLP, the diagnosis can be performed in this family.
- *Family 2.* By a reasoning previously described (family 1), it is possible to conclude that the affected child has genotype (m,+)/(m,+) and that the parents are respectively (N,−)/(m,+) and (N,+)/(m,+).

  A problem of ambiguity is raised in the mother who is homozygous for the RFLP which means that the transmitted allele '+' can be both the one linked with the N and the one linked with m. If the father transmits the RFLP '−', the diagnosis is not a problem because the fetus will not be affected whether he is N/N or N/m.

  Here the father transmits the allele '+' and by co-transmission, the allele m, therefore, the child will be, depending on the maternal input, impossible to identify m/m affected or m/N healthy carrier, which re-quires parallel analysis (using another RFLP that would be more informative, or even better, a microsatellite or a VNTR, that is almost always informative, see above).
- *Family 3.* By a reasoning previously describing (family 1), it is possible to conclude that the affected child has the genotype (m,−)/(m,+), and that the parents are (N,−)/(m,+) and (N,+)/(m,−) without knowing which is the father's and which the mother's. The genetic schema also leads to ambiguous situtations: if the child to be born is +/+ or −/−, it is, considering the parental chromosome, a healthy carrier N/m, without knowing if the mutation is of maternal or paternal origin; if the child is +/−, then he can be N/N or m/m with the same probability.

  Here again, an analysis with another RFLP or VNTR can remove the ambiguity, or even a simple ASO analysis. Indeed, if one parent carries a mutation detectable by ASO or ARMS or OLA and one of these techniques has identified the presence of this mutation in the fetus, the fetus can only be m/m if he is +/−.

Figure 1.22   Family analysis of the co-transmission of an RFLP with a pathological mutation

Application for the diagnosis of loss of genetic material

Analysis of RFLP polymorphisms is also applied in the study of loss of heterozygosity associated with deletion disomies or recombinations, notably in the macromutations often affecting tumour cells (see Section 3.3.3)

Identification of gene mutations using PCR and RFLP

Some mutations can generate or remove an RFLP site. It is therefore easy to identify their presence by PCR amplification of the sequence of interest containing the mutation whose presence or absence has to be tested. After amplification, the amplimers are submitted to the action of an endonuclease associated with the site created or removed by the mutation; visualization on a gel of the obtained fragments will allow a direct deduction of the presence or the absence of the mutation. This PCR-restriction approach is simple and is often used when a mutation creates or removes a restriction site (haemochromatosis or haemoglobinopathy.

In the case where a mutation does not create or abolish an RFLP site, it is possible to create one artificially using a primer not strictly complementary (for one single well chosen base) but able to hybridize in a first cycle performed with an annealing step at low stringency. For this purpose, one of the two amplification primers has a sequence defined to hybridize just before the mutation site and contains a base that can generate an RFLP site on the amplimer in the case that the tested site is mutated, while no RFLP site will be created at the test site if not mutated. The following cycles can be performed with a normal stringency during the annealing steps. The PCR will be followed by a restriction and visualization on a gel where the presence of fragments reveals the presence of the mutation (the ARMS technique is a variant of this technique)

1.9.9   VNTRs and microsatellites

Variable number of tandem repeats (VNTRs), also called 'minisatellites', are a very different type of polymorphic molecular marker than RFLPs. They consist of short DNA sequences, most of the time dinucleotides, in tandem repeats. The microsatellites are tandem repeats of two, three or four nucleotides. The best known and most used are the CA and TA repeats. These repeats are very frequent and cover the genomic DNA.

At one given repeat locus, the number of repeats can vary so that one individual has a very small chance of having the same number of repeats at two homologous sites.

Considering the important number of alleles for a repeat at one particular site (as many alleles as different values in the number of repeats), the number of possible genotypes is very large (see Chapter 7), much larger than the three possible genotypes

Box 1.14   Microsatellite DNA are very polymorphic genetic markers

A microsatellite sequence is made up of a tandem repeat of two, three or four nucleotides (di-nucleotide repeat, tri-nucleotide repeat, tetra-nucleotide repeat) at a precise position in the genome. A microsatellite sequence can form a genetic marker if the number of repeats vary at this position in a way that all individuals, being diploid, have a high chance not only of being heterozygous, but also to have very different genotypes, especially if there are a large number of alleles for this microsatellite. These microsatellite sequences are part of the favourite genetic markers for genetic fingerprints or for some types of genetic diagnosis tests, notably the prenatal tests.

for an RFLP site. Therefore, two individuals have little chance of sharing the same genotype for a VNTR marker or microsatellite and a probability almost null to be genetically identical for several; this is why these polymorphisms are at the basis of the genetic fingerprints techniques that allow, with great reliability, the analysis of a biological sample (blood or sperm) to exclude or identify an individual supposed to be at the origin of this material (paternity testing, legal identity, criminal enquiries, see Chapter 6). This is also why microsatellites tend to replace the RFLP in indirect diagnosis (see haemophilia, Figure 2.17).

On the technical side, microsatellite studies were first performed using the Southern method but are now almost exclusively done using PCR. The fragment containing the microsatellite site is amplified using labelled primers, then the PCR products are submitted to electrophoresis together with a molecular marker, in order to determine the size of the fragment when it passes in front of the laser window at the bottom of the gel or of the capillary, and to genotype the individual or biological sample tested. It should be noted that the size of the repeat is sometimes such that PCR cannot be performed and the Southern method is the only way to study the repeat sequence (see fragile X syndrome).

The analysis of the genetic diversity of a repeat sequence is not only useful in the application of forensic medicine (genetic fingerprints), but also constitutes the basis for the genetic diagnosis of some hereditary diseases where the pathogenic mutation is an amplification of a repeat sequence (see fragile X syndrome, neurodegenerative triplet diseases), or in certain types of cancer such as *hereditary non-polyposis colorectal cancer* (HNPCC) in which tumour cells are characterized by a replicative instability leading to a large variation of the number of repeats in the tumour clone.

1.9.10   *Single nucleotide polymorphism markers*

Single nucleotide polymorphism (SNP) markers are di-allelic polymorphisms concerning a base pair but not necessarily leading to the formation of the suppression of a restriction site, as is the case with the RFLP markers. They are distributed all along the genome, within or outside of coding sequences.

The increasing number of identified SNPs constitutes a genetic analysis tool rivalling RFLP, VNTR or microsatellites, notably for indirect genetic diagnosis. They can be tested by various techniques like ASO or the techniques derived from ARMS, but mainly by OLA or DHPLC.

1.9.11   DNA microarrays

The miniaturization that touched a technology like informatics has developed even faster in biotechnology. Indeed, the molecular hybridization tests between a target sequence and a test probe that rely on optical visualization (Southern, Northern, dot or reverse-dot blot, see Sections 1.9.5 and 1.9.6) can be performed nowadays on a large scale thanks to the nano-technology in DNA microarrays.

These consist of a glass or a silicon surface of 1.6 cm^2 divided into a very large number of quadrants, on the surface of which are oligonucleotides capable of hybridizing with a population of genomic DNAs or cDNAs. A DNA chip functions like a reverse dot (see Section 1.9.6) but the oligoprobe density is such that thousands of different molecules can be tested simultaneously and can even be quantified depending on the intensity of the signal collected by the visualization system (computer controlled confocal microscopy). The two main applications of DNA chips are transcriptome studies and identification of genetic mutations.

Transcriptome studies

The transcriptome is the equivalent of a reverse Northern blot and allows the qualitative and most of all the quantitative analysis of the level of gene expression in a particular tissue at a particular time. For this purpose the microchip is divided into many spots, separated by about 400 µm, and on each spot is a given quantity of a gene specific probe, cDNA or synthetic oligo-probe. The microarray is then hybridized with a cDNA library, made by RT–PCR on the mRNA collected from a tissue sample, and labelled with a fluorochrome. This allows quantification of the expression of all genes present on the chip in the tested tissue, that is to say, its 'transcriptome', through analysis of the fluorescent signals (Figure 1.23)

The microchip density is such that 260 000 oligonucleotides, corresponding to the testing of the expression of 6000 yeast genes, have been laid on a surface of only 1.28 cm^2 (Figure 1.24) and that the expression of the 8000 genes making up the *Arabidopsis thaliana* genome can easily be laid on a single microchip in order to test their level of expression in the stem, leaves, root or flower tissues, giving the transcriptome of each tissue.

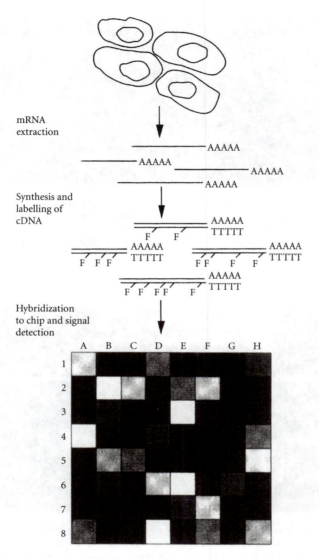

mRNA
extraction

Synthesis and
labelling of
cDNA

Hybridization
to chip and signal
detection

Use of chips carrying cDNA or oligonucleotides. Messenger RNAs are extracted from a tissue or cell culture, fluorescent complementary DNA is synthesized using labelled nucleotides (F), the messenger RNA is then eliminated, and the cDNA is hybridized to a chip containing as array of cDNA molecules or oligonucleotides. Detection allows the evaluation of the presence and abundance of each messenger RNA: for example, the sequences matching B2 or E3 are abundant, those matching A1 or C2 are less frequent, and none match B1 or F1.

Figure 1.23   Protocol to set up a tissue transcriptome

In humans, the transcriptome is being used systematically to analyse the differences in the expression levels of genes in normal cells compared with tumour cells and to evaluate the effect of anti-tumour treatments (Figure 1.25).

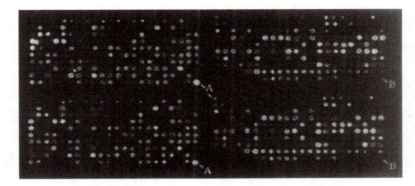

Hybridization of cDNA onto chips. cDNA representing 95 per cent of yeast coding sequences have been fixed onto a slide. The spots are seperated by 380 μm. The positive controls (A) represent the complete yeast genome, the negative controls (B) are phage DNA. Shown on this figure are two columns of different matches each comprising 219 distinct cDNAs. Each quadrant is vertically duplicated, to control for the reliability of the hybridizations obtained.

Figure 1.24   Yeast transcriptome

Hybridization on an oligonucleotide chip. Part of the oligonucleotide chip hybridization allowing the analysis of 8102 genes is shown. On the horizontal axis are oligonucleotides allowing the detection of transcription of some of these genes (group f: γ immunoglobulin, CSF (colony stimulating factor), NSF (neutrophil cytosolic factor), seven regions of the λ immunoglobulin chain, a class II MHC protein, μ immunoglobulin, EDR2 (early development regulator 2), MIP-1 (max-interacting protein 1); group g: homeobox-1, somatomedin C, KIP-2 (cyclin dependent kinase inhibitor P57), five different proteins interacting with fatty acids, two collagen receptors (CD36), a glutathione peroxidase, LIM protein, an alcohol dehydrogenase, a glycerol-phosphate dehydrogenase, a protein recognizing retinol, a lipase, α7 integrin, four channels of the aquaporin type, A1 apolipoprotein, a cytokinin, and two receptors for endothelin). Each oligonucleotide occupies a 24 × 24 μm surface. These chips have been hybridized to cDNA originating from 84 breast cancer patients, before and after treatment with doxorubin (vertical axis).

Figure 1.25   Transcriptome, on oligonucleotide microarrays, of breast cells from 84 patients suffering from a breast cancer, before and after treatment with the doxorubine – the lighter the spot the higher the fluorescence signal, and therefore the higher the expression level of the tested gene

Identification of gene mutations

The first technologies for the identification of gene mutations were specific for a particular mutation (see DF508 for cystic fibrosis, or C282Y for haemochromatosis) but the identification of a large number of mutations in a single gene made it necessary, for reasons of both time and cost, to develop protocols capable of testing simultaneously for the presence of several mutations in the same gene (see Section 1.9.6, ASO, or Section 1.9.7 ARMS, OLA).

However, the most efficient of these methods can only test 10 to 30 mutations in one gene. With the oligo-chips, the efficiency will be incomparable because it should be possible to perform 'genome scans' meaning the identification of all the possible point mutations on the whole coding sequence of a gene, its intron/exon boundaries, its promoter and terminator sequence. It just happens that DHPLC technology (Figure 1.10.4) is still the most affordable on the economical aspect.

In contrast, when its cost becomes affordable, DNA microarray technology will become very useful for the systematic screening of at risk couples for a recessive disease, before the birth of a first affected child (see cystic fibrosis Section 2.2.1 and Section 7.4). Indeed, microarrays carrying oligonucleotide probes of all known mutations of a gene (less than all possible mutations) could be used to screen the healthy carriers of these mutations and be advise about the possibilities of prenatal screening for the first pregnancy.

It is evident that such a technology, so efficient for allelic screening, could used for purposes less laudable than severe diseases prevention; as for any new technology, the DNA microarray can be misused by those who wish to do so, if such uses are not controlled.

1.10   Other techniques to study allelic diversity

1.10.1   Introduction

In some circumstances, genetic diagnosis is difficult or impossible because the gene studied is mutated but carries a rare mutation, undetected in a routine test, or even a mutation as yet unidentified.

The establishment of the diagnosis requires screening strategies, that is to say a systematic search for a mutation along the entire gene sequence. This search can be difficult due to the size of the gene, notably if it has many large introns. When this is not the case, it is possible to amplify the exons by PCR and to sequence them, bearing in mind that PCR fragment sequencing has now become a simple, reliable and not that expensive routine. However, some mutations can be missed during the sequential analysis of the exons (intronic mutations or mutations in the promoter or the 3' end of the gene).

Box 1.15   Heteroduplex analyses are alternative tools for detecting DNA polymorphisms

Genetic mutations and, more generally, polymorphisms carrying a single nucleotide change can be identified by the fact that the hybridization between the wild type DNA strand and the mutated one leads to a local mismatch that can be observed using an adapted apparatus, avoiding the cost and the time required by older techniques such as Southern blot, dot blot or reverse dot. These analyses are: single-strand conformation polymorphism (SSCP), denaturing gradient gel electrophoresis (DGGE) and DNA high-performance liquid chromatography (DHPLC), in order of increasing usage and efficiency.

When the sequencing is not possible other more appropriate strategies have been developed by researchers, with four being quite frequently used: single-strand conformation polymorphism (SSCP), denaturing gradient gel electrophoresis (DGGE), DNA high-performance liquid chromatography (DHPLC) and the protein truncation test (PTT) whose purpose is to identify a mutation/polymorphism in the gene sequence without requiring systematic sequencing.

1.10.2   Single-strand conformation polymorphism (SSCP)

This consists of amplifying many fragments of the studied gene using PCR, then denaturing them before cooling them very rapidly in order to prevent renaturing, leading to the folding of the single strands in a conformation specific of their respective sequence (at least, that is what is expected) during a non-denaturing gel electrophoresis.

If the two copies of a gene are identical, all the PCR of the different fragments of the gene will contain only one type of double-stranded DNA fragment and two types of single-stranded DNA, appearing on the gel as two different bands if their conformation leads to a different migration speed.

If the two copies of a gene differ for a point mutation, one of the PCRs – the one involving the segment carrying the mutation – will contain two types of double-strand DNA fragments, one normal and the other one mutated, and four types of single-strand fragment, appearing on the gel as four different bands if their conformation leads to a different migration speed.

This method is simple and widely used despite the fact that many mutations stay undetectable (detection level is about 70 per cent) because they do not have a large enough effect on the conformation of the mutated strands for their migration to be distinguishable from the non-mutated strands.

Finally, it is important to note that this method allows the identification of point mutations of the nucleotide sequence, and that it is necessary to check if it has a pathogenic effect, i.e. if this polymorphism is the suspected pathogenic mutation.

1.10.3   Denaturing gradient gel electrophoresis (DGGE)

This method is much more efficient (a detection rate of up to 95 per cent) than the SSCP one but it is much more complicated to set up. The principle of the method relies on the observation that the melting (or denaturing) temperature of a double-stranded fragment depends on its sequence in a constant environment, and on the environment at a constant temperature. Two DNA fragment with a single base pair difference can have very different melting conditions.

The double-stranded fragments obtained by PCR are loaded on a gel where the de-naturing conditions are increasing (increase in urea and formamide concentrations). Two fragments with a single base pair difference will not denature at the same level in such a gel. As soon as a fragment is denatured, its migration is severely slowed down because the two single-stranded DNA are still bound to each other thanks to a Psoralene molecule added at their extremities after the PCR, which suddenly increases its volume and therefore its resistance to passing through the gel matrix.

If two copies of a gene are identical, all PCRs will contain a single type of double-stranded DNA fragment, leading to a single band on a denaturing gel. If the two copies of a gene are mutated and carry the same mutation, all PCRs will contain one single type of mutated double-stranded DNA fragment leading to a single band but at a level that can be different from the level reached by the non-mutated fragment. If the two copies of a gene differ for a point mutation, one of the PCRs of the gene, involving the segment carrying the mutations will contain two types of fragments called a 'heteroduplex' made up of one normal + strand renatured with a mutated − strand and a normal − strand renaturated with a mutated + strand (with a mismatch at the level of the mutation). Indeed, these heteroduplexes form spontaneously during PCR and a final additive step can even favour their formation, which is interesting for the analysis because their presence confirms that a polymorphism (a mutation) is present. The DGGE can show up to four bands corresponding to four types of fragments if the melting conditions differ (Figure 2.4).

The gradient of denaturing agent concentration should be adapted to the sequence of the studied fragments, the standard fragment being the reference to determine the denaturing conditions. As is the case for the SSCP, it must be noted that this method reveals a polymorphism in the nucleotide sequence, but it must be tested to check if it is the suspected pathogenic mutation.

1.10.4   Searching for polymorphisms using DHPLC

DNA high-performance liquid chromatography (DHPLC) is liquid chromatography adapted for the identification of SNP type polymorphisms. This display, even if it is different from the one used for DGGE (see above), is based on the same principle, which consists of the identification of heteroduplex molecules obtained by PCR after simultaneous amplification of the two different allelic sequences from a heterozygote. It has the same reliability as DGGE.

The PCR products are loaded on a polystyrene column and then eluted with an acetonitrile gradient buffer. The DNA fragments will be easier to elute the lower their melting temperatures, this temperature depending on the duplex size and its GC/TA composition but, most of all, on a possible mismatch in the case of a heteroduplex.

Knowing the standard elution profile of the sequence obtained in a standard homozygote, it is then easy to notice the variation in elution characteristics of the presence of an SNP type mutation that can be identified by sequencing afterwards.

For SNPs which are already known, reference elution profiles for the carriers are available, which allows the direct identification of their presence (see oncology applications, Chapter 3).

1.10.5   Protein truncation test (PTT)

This test is designed to search for a mutation in a gene by finding it directly in the protein encoded by this gene, when the consequence of the mutation is essentially the shortening of the peptide sequence (stop mutation, frameshift mutation, partial deletion, mainly in the very big genes).

The method (Figure 1.26) consists of RT–PCR with a − primer that will allow the formation of a − strand using reverse transcriptase, and a + primer carrying at the 5′ end an extra sequence corresponding to the phage T7 promoter, in a way

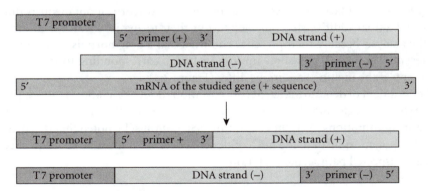

- The mRNA of the target gene is submitted to RT–PCR with a (−) primer in order to make a (−) strand.
- The (+) primer allowing the synthesis of a (+) strand carries the T7 phage promoter upstream.
- Fragments of the gene defined between the (+) and (−) primers fused to the T7 promoter are collected, allowing a coupled *in vitro* transcription/translation in the presence of tRNA filled with radioactive amino acids.
- The synthesized peptides are separated by PAGE–SDS electrophoresis and transferred onto a membrane.
- The Western blot is auto-radiographed and the size of the fragments is determined by comparison with the standard.

Figure 1.26   Principle of the protein truncation test (PTT)

that the DNA fragments amplified by RT–PCR can be *in vitro* transcribed (the T7 sequence at the 5′ of the primer does not affect its capability to anneal with the − strand synthesized from the − primer, Figure 1.19).

The fragments obtained are transferred to an *in vitro* transcription/translation system (reticulocytes lysate) in the presence of tRNA, some being labelled with radioactive amino acids. After peptide synthesis, the content of the tubes is loaded on an electrophoresis gel in the presence of SDS, a denaturing agent that will separate the peptides according to their size. After transfer of the peptides onto a membrane, the *Western blot* is auto-radiographed and the size of the fragments is determined by comparison with the size of the standard fragments, obtained from the standard gene that co-migrated at the same time

This method is widely used for the search of mutations in the gene *APC* (*Adenomatous Polyposis Coli*) involved in cancers with predisposition for colon polyps, or the genes *MSH2* and *MLH1* involved in cancers without predisposition for colon polyps and most of all, the gene for dystrophin involved in Duchenne and Becker dystrophy (Section 2.3.3).

2 The diagnosis of inherited diseases

Catherine Boileau, *CHU Paris Véronique David,*
CHU Pontchaillou
Emmanuelle Girodon, *CHU Henri-Mondor*
Éric Le Guern, *CHU Pitié-Salpétrière*
Étienne Mornet, *Université de Versailles*
Véronique Pingaud, *CHU Henri-Mondor*
Serge Pissard, *CHU Henri-Mondor*
Jean-Louis Serre, *Université de Versailles*

2.1   Introduction

The increase in the number of genes cloned and the identification of the mutations in those genes responsible for a disease, both contribute to the progress made in human genetics and in genetics in general, both in basic research and in its applications. This chapter deals with the applications that can be made from this research, but it is important to present first the major input introduced by gene cloning and sequencing.

The identification of the mutations responsible for a particular disease gave rise to a new field: molecular pathology. It consists of linking the mutation with its biochemical effects on the gene product, quantitatively or qualitatively. It helps to explain the clinical observations due to this biochemical effect, in one particular cell type or at one particular developmental stage. Therefore, 'molecular pathology' contributed extensively to the elucidation of complex links between the clinical phenotype, the disease and the pathological genotype, sometimes leading to unexpected results.

Formally, a hereditary disease is Mendelian when it is caused by a single gene. A disease is recessive if the patient carries two mutated alleles (copies) of the gene, both parents being heterozygotes, with one functional allele and one pathological

Diagnostic techniques in genetics   Edited by Jean-Louis Serre
© 2006 John Wiley & Sons, Ltd

allele, especially if the disease causes death before reproductive age. Functionally, the 'normal' allele is considered to be the one that compensates for the effect of the pathological allele in the healthy individual. This shows why two mutated copies are necessary to cause the disease.

A disease is dominant if one mutated copy of the gene seems sufficient to cause the disease. In this case, the 'normal' allele activity cannot compensate, or compensates only partially, for the effect of the mutated allele, showing why one mutated copy is sufficient to cause the disease.

However, this functional link is very theoretical and molecular pathology is indispensable in the full explanation of the links between the genotype (which gene, which mutation, which effect) and the phenotype (which disease, which clinical form).

2.1.1 Different mutation classes

A gene cannot be restricted to a coding sequence because it also contains sequences necessary for its expression, i.e. the promoter, the 5′ and 3′ untranslated regions (5′UTR and 3′UTR), the polyadenylation signal, and intronic sequences that have to be very precisely excised to reconstitute the exact coding sequence of the messenger RNA (or several coding sequences if the gene is alternatively spliced). Therefore, a mutation in a gene will have different effects depending on its site and its nature.

In the end, at the functional level, we distinguish two types of mutation depending on whether they affect the quantity or the quality of the gene product. In the latter case, the mutation generally affects the coding sequence.

The mutations can also be divided at the functional level into two other categories, not overlapping the two previous ones.

- The loss of function mutations causes a decrease or a loss of the gene product (quantitative mutations) or a decrease or a loss of the activity of the gene product (qualitative mutations);

- The gain of function mutations causes an increase in the amount of gene product (quantitative mutation) or an increase in its activity, and sometimes creates a new property, leading to a toxic product responsible for a pathological effect.

Depending on the mutated region in the gene (promoter, intron, coding sequence, poly(A) site, 5′ or 3′ UTR) and depending on its type (base pair substitution, deletion or insertion of one or several base pairs, repeated sequence extension), a mutation will have a quantitative or a qualitative effect leading to a loss or a gain of function. It is important to mention that all combinations (site, nature, effect) are possible but some are more often seen than others. A stop mutation in the coding sequence usually leads to a loss of function but might occasionally cause a gain of function.

Box 2.1   A mutation in one gene may lead to a loss or a gain of function of this gene

- When it has an effect, a mutation in one gene always leads to a loss or a gain of function.

- The loss of function can be the result of a quantitative effect of the mutation – there is less of the product (partial loss of function) or there is no more product (complete loss of function); it can also be the result of a qualitative effect – the product is less functional or not functional at all.

- The gain of function can be the result of a quantitative effect (there is more product) or qualitative effect (the product is more active); it can also correspond to the emergence of a new property or biological activity.

2.1.2   Dominance and recessivity are explained by molecular pathology

Loss of function mutations frequently have a recessive effect compared with the functional gene effect, the latter having an activity sufficient to compensate for the loss of the mutated gene. This is the case for most of the recessive diseases (β–thalassemia, cystic fibrosis, haemophilia, Duchenne/Becker myopathy).

On the other hand, many dominant pathologies are not caused by loss of function mutations but are caused by gain of function mutations, also referred to as dominant negative mutations when they lead to the presence of a toxic product (haemolytic anaemia with unstable haemoglobin, Huntington's disease and spinocerebella ataxia where the CAG triplet extension leads to the formation of an expanded and toxic poly-glutamine domain).

Box 2.2   Dominant versus recessive diseases

- A disease is called recessive when it requires the simultaneous presence of two mutated copies of the gene involved in that disease. In contrast, when only one copy of the gene is sufficient, despite the presence of a wild type copy, the disease is called dominant.

- As a consequence, for a recessive disease, the heterozygote that carries a mutated and a functional copy of the gene is not affected – he is called a healthy carrier. For a dominant disease, the carrier of two mutated copies of the gene is affected *a fortiori* because one copy is sufficient to be affected. In fact the carriers of two mutated copies are very rare and the effects are often more serious, even sometimes lethal.

The associations 'loss of function/recessivity' and 'gain of function/dominance' are too frequently made; but even if the correlation is strong, it is not absolute, and famous counter-examples exist.

- Some dominant diseases are caused by loss of function mutations and can only be partially compensated for by the functional allele activity (dosage effect), the cell or the tissue needing a minimum amount of product superior to the 50 per cent present. The loss of function mutation is called dominant by haplo-insufficiency (familial hypercholesterolemia, mature-onset diabetes of the young (MODY)). This 'dosage effect' is often illustrated by the fact that the severity of the disease is linked to the intensity of the loss of the function and that the patients carrying two mutated copies are more severely affected.

- In contrast, there are some recessive diseases associated with gain of function mutations, like sickle-cell anemia, caused by haemoglobin HbS, whose polymerization capacity is toxic in the homozygote, but only mildly deleterious in the healthy carriers because this haemoglobin is then diluted.

- Finally, there are some paradoxical special cases, like fragile X syndrome (see below) and hereditary retinoblastoma, where a loss of function mutation leads to a dominant disease, not because it is haplo-insufficient, but because of a biological peculiarity. In the fragile X syndrome (Section 2.3.1) this peculiarity is the random inactivation of one of the X chromosomes in females. In the hereditary retinoblastoma, only a few cells in the organism are affected. Indeed, the affected parent transmits a loss of function mutation in the *RB* gene, completely compensated for by the retina cells. However, because the number of mitoses during eye embryogenesis is very important, some cells lose the other copy by somatic mutation and therefore become tumourous. This phenomenon is certain enough to affect several independent cells, at the origin of multiple tumoural 'nodules' in both eyes. This is

---

Box 2.3   Loss and gain of function mutations may be either dominant or recessive

- A loss of function mutation can lead to a recessive disease if the heterozygote is a healthy carrier because a single functional allele, not mutated, is sufficient to give the physiological function of the gene (haplo-sufficiency), but can lead to a dominant disease if the functional allele is not sufficient (haplo-insufficiency).

- A gain of function mutation can lead to a dominant disease if the mutated allele has a toxic effect that cannot be counteracted by the effect of the not-mutated allele (dominant negative effect) but can lead to a recessive disease if this toxic effect has to be homozygous to give a pathology with a detectable phenotype.

a form of hereditary cancer transmitted in a dominant fashion (each parent has half of its descendants affected) although the mutation responsible for the pathology is a loss of function completely compensable for (recessivity at the level of the cell).

Therefore molecular pathology can clarify the dominance or the recessivity of Mendelian diseases, in defining the causal relationship between the genotype and the pathological phenotype.

2.1.3 Genetic heterogeneity can be explained by molecular pathology

Genetic heterogeneity means that one phenotype or one disease can correspond to various genetic situations and, more generally, that the relationship between genotype and phenotype is often more complex than expected.

Inter-locus heterogeneity: one disease/several genes

The simplest way to consider a Mendelian disease is as a monogenic disease, meaning a disease for which every patient is affected in the same gene (one disease/one gene). This is the case for diseases like phenylketonuria and cystic fibrosis.

However, even before molecular biology existed, the fact that one particular disease could be due to mutations in different genes, even if each patient is mutated in only one of those genes, was well understood. The disease is *monofactorial* because each patient is only mutated in one gene, but heterogeneous (not monogenic) because this gene can be different from one patient to another. This is the case, for example, for haemophilia, where you can distinguish patients with a deficiency in factor VIII (haemophilia A) from those with a deficiency in factor IX (haemophilia B). This distinction is based on the biochemical properties of the two disease-causing products. In other circumstances (Charcot–Marie–Tooth), genetic heterogeneity has been shown by the fact that one disease could be transmitted as an autosomal disease in some families, while it could be sex linked in others, meaning that at least two genes are involved, one autosomal and the other on the X chromosome. Genetic heterogeneity can also be shown when all the descendants of two affected parents are healthy (for example in the case of monofactorial recessive deafness). This result is expected after functional complementation when two mutations are carried by different alleles (affecting different genes). With the localization, cloning and sequencing of the genes, molecular biology helped confirm the heterogeneity of some diseases (Charcot–Marie–Tooth, Fanconi anaemia, *Xeroderma Pigmentosum*) and helped discover that some diseases thought to be monogenic were in fact heterogeneous and simply monofactorial (diabetes MODY with at least four autosomal genes, pigment retinitis with eight autosomal genes for seven dominant forms and one recessive and three genes on the X chromosome).

Intra-locus heterogeneity: one gene/several diseases

A more unexpected result in molecular pathology is shown by another kind of heterogeneity that associates several diseases to one single gene instead of associating one disease to several genes. This situation, already known in the case of mutations in the genes encoding for haemoglobin α or β that are responsible for several different forms of haemoglobinopathy (see below), appears to be more frequent than was first thought. For example, Duchenne and Becker myopathies have different clinical characteristics (see below) and are due to different mutations in the gene encoding for dystrophin. Four sometimes very different diseases like Hirshsprung disease (malfunctioning of the colon due to a lack of innervation) and three types of hereditary cancers (multiple endocrine neoplasia type 2A or 2B and medullary thyroid carcinoma) are the result of mutations in the RET gene, affecting different domains of the protein and leading to dominant negative loss of function (Hirshsprung) or gain of function (cancers) depending on the case. Likewise, when the PMP22 gene is duplicated in tandem on chromosome 17 (three copies in the patient), it is responsible for one of the severe adult forms of Charcot–Marie–Tooth (CMT 1A) disease, while its deletion (only one copy on the patient) leads to a benign hereditary neuropathy (painful when the nerves are pressed). In each case, the disease is dominantly transmitted and follows, in half of the offspring, the transmission of the mutated allele (duplication or deletion).

The existence of several diseases linked with mutations in the same gene is only one extreme form of intra-locus or allelic heterogeneity. In most cases, allelic heterogeneity is only responsible for part of the variability in disease expression, meaning the existence of forms more or less serious with more or less clinical signs (see Section 2.2.1, cystic fibrosis). It is worth mentioning that another part of the variability in disease expression is the result of the action of other genes, the modifier genes, different from the gene principally involved in the pathology, that reduce the severity of the pathology (fetal haemoglobin still present in thalassaemia or drepanocytosis) or increase it (meconium ileus in 15 per cent of newborns affected with cystic fibrosis). This diversity of disease expression due to modifier genes can explain the clinical heterogeneity between families or populations as well as the heterogeneity between affected patients within the same family.

Because it permits the determination of all causal effects, from the primary defect within the gene to the different clinical signs of a disease, human molecular pathology has lead to a therapeutic way of thinking and doing research. It has also allowed the whole genetic field to progress in understanding the genetic message and the consequences of all types of mutation in the message. Indeed, doctors and hospitals have played an exceptional role in screening for 'natural mutants' in humans affected in essential genes for cellular functions, those being impossible to detect in other species or to induce in research animal models like mice or drosophila. Nowadays,

---

Box 2.4   Clinical and genetic heterogeneity

- Many diseases present a clinical heterogeneity (phenotypic heterogeneity) characterized by the severity or the extent of the symptoms, by the severity or the way the disorder evolves.

- One part of the clinical heterogeneity can be explained by allelic heterogeneity, by the fact that different mutations can have variable phenotypic effects, either in their type or in their strength.

- Another part of the clinical heterogeneity, notably between affected individuals within the same family, can be explained by the effect of modifying genes that can increase or decrease the effect of pathological mutations in the gene principally involved, without being responsible for the appearance of the pathology.

- Finally, the effect of the environment on the appearance of the disease and on its clinical variability should not be neglected.

---

molecular pathology can pursue its fundamental studies on animal models and apply the results to humans, so far through genetic diagnosis, as the next chapter will show.

2.2   Example diagnoses for autosomal diseases

2.2.1   Cystic fibrosis

Introduction

Cystic fibrosis, also called pancreatic cystic fibrosis, is the most frequent of the severe recessive autosomal disorders in the Caucasian population. About one newborn out of 2500 is affected (between 1/1600 and 1/3500 in France), which corresponds to a carrier frequency of 1/25 (see Chapter 7).

In its classical form, it is a paediatric disease associated with both lung and digestive problems. However, the disease expression varies with respect to the age of onset, the diagnosis circumstances and the severity of the disease. Some forms appear just before or at birth, or during the first few months in the form of severe bronchial infections, but also later during childhood, adolescence or even adulthood. Symptoms can be more digestive or more respiratory. Also, for a certain number of cases for which clinical and/or biological diagnosis is not obvious, genetic studies happened to be especially effective. The characterization of the mutations in the cystic fibrosis trans-membrane conductance regulator (CFTR) gene allows linking cystic fibrosis to some adult symptoms such as male sterility and some types of diffuse bronchiectasis chronic pancreatitis, showing a real continuum in the expression of the CFTR gene mutations. Since the gene was cloned in 1989, over 1000 mutations have been described. However, for most of them, the deleterious effect can not be easily proven because the functional tests (cellular mechanisms leading to maturation and membrane expression of the mutant protein) are still part of the research process. Correlation studies between genotype and clinical phenotype are indeed difficult, especially since the mutation diversity can only explain a part of the variability of the clinical phenotype. It is therefore difficult to propose a prognosis when considering a given genotype. Despite those difficulties, and considering the genetic counselling aims, molecular studies of the CFTR gene are important, and many laboratories are now performing such studies using a wide range of tools.

Clinical description

Cystic fibrosis was first described in 1938 by Andersen as a pancreatic cystic fibrosis, a clinical pathology responsible for malabsorption and growth retardation hardly dissociable from other causes of malabsorption. The pulmonary attack, due to recurring infections, started to be considered as a major factor for diagnosis in the 1940s. The observation made by diSant'Agnese, during a heatwave in the United States in 1953, that children suffering from cystic fibrosis tend to become dehydrated faster than the others, with the excess of salt in their sweat, lead to the development of a reliable biological test, the sweat test, which was the major diagnostic criterion up until the discovery of the gene. This test reveals an increase in chloride and sodium ions in the sweat of affected patients. Most of the lesions of the target organs can be explained by the obstruction of the conduits due to secretions or to thick, dehydrated mucus: the recurring bronchi and sinus infections; gut obstructions (about 15 per cent of the children show such an obstruction at birth, the meconium ileus); an exocrine pancreatic insufficiency responsible for steatorrhea (oily stool) due to malabsorption of fat, found in 85 per cent of the children; a hepatic attack that can lead to cirrhosis in 5 per cent of cases; a progressive atresia of the vas deferens in boys, responsible for infertility in 98 per cent of the cases. Identification of mutations leads to a revision of the diagnostic criteria. Even if a positive sweat test stays an essential criterion for diagnosis in most cystic fibrosis cases, it can fail and certain pathological conditions, like severe malnutrition, can lead to a positive sweat test.

Identification of mutations in the CFTR gene allows the recognition of some non-typical forms of the disease with a late onset or with mainly respiratory or digestive symptoms. More recently, it has allowed the expansion of the spectrum of diseases involving the CFTR gene to monosymptomatic forms in the adult, like male infertility due to a lack of vas deferens, or certain types of diffuse bronchiectasis (bilateral dilatation of the bronchi) or chronic pancreatitis. The distinction between the concepts of 'disease spectrum' and 'continuum' in cystic fibrosis is fairly small.

From the CFTR gene to the protein

The gene responsible was characterized in 1989 using a positional cloning approach, by the North American teams of L.C. Tsui and F. Collins. It is a large gene located in region 7q31 on the long arm of the chromosome (Figure 2.1). It extends over 250 kilobases (kb) of DNA and contains 27 exons. It is transcribed as a 6.5kb messenger RNA and it encodes for a transmembrane protein with 1480 amino acids.

The CFTR protein is expressed at the apical pole of the epithelial cells of the pancreatic, biliary and intestinal tracts, tracheobronchial tree, renal tubules, genital ducts and sweat glands. The level of expression varies depending on the tissue and the developmental stage. Environmental factors and the biology of the respiratory epithelial cells probably play a determining role in the physiopathology of cystic fibrosis, more than the level of expression of the CFTR protein. For example, the lack of any kidney pathology despite the fact that the protein is expressed in the kidney ducts suggests that other tracts can compensate for the hydroelectric transport defect due to the absence of functional CFTR in affected patients. In contrast, the CFTR protein is only mildly expressed in the respiratory cells even though the respiratory tract is the main target in cystic fibrosis. The anatomical structure is also an important element: The narrow vasa deferentia are especially sensitive to dehydration of the intraluminal fluid.

There are several functions for the CFTR protein. It is a chloride ion channel (Cl^-) that makes Cl^- ions exit the epithelial cell. Because this channel has a defect in cystic fibrosis, the Cl^- ions are kept in the cells and block the water from freely exiting, leading to dehydration of the secretions and mucus. At the level of the sweat glands, the ions transfer in the reverse direction, which explains the elevated concentration of Cl^- ions in the sweat of affected patients. Its involvement in regulating other ion channel functions explains how the hydroelectric transport disequilibrium observed in cystic fibrosis increases mucus dehydration, which in turn is mainly responsible for obstructive phenomena in the different target organs. In addition, the high frequency of respiratory infections due to Pseudomonas aruginosa (bacille pyocyanique) in patients suffering from cystic fibrosis suggests that a defect in the CFTR protein could be directly or indirectly linked with this susceptibility.

Molecular pathology and genotype–phenotype correlations

Molecular epidemiology As soon as the gene was discovered in 1989, the most frequent mutation, named ΔF508 because it corresponds to the loss of a phenylalanine

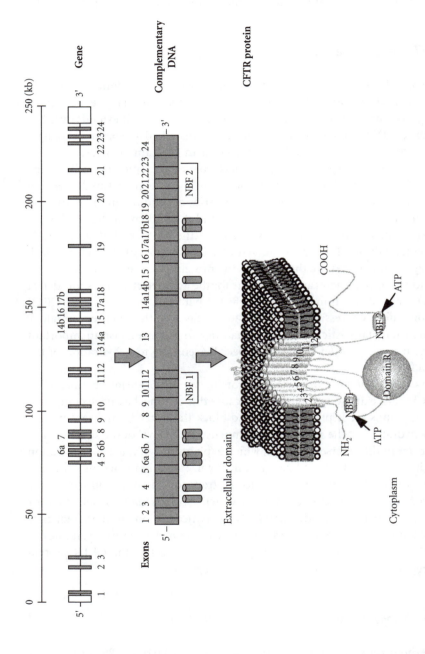

Gene: the grey rectangles indicate the exons, the white indicate the sequences transcribed but not translated.
Complementary DNA: the cylinders represent the sequence coding for the transmembrane domains of the protein.
NBF1 and 2: nucleotide binding fold (ATP binding domain).

Figure 2.1   From the gene to the protein

residue at position 508 of the protein, was identified. This mutation is found in about 70 per cent of the mutated chromosomes, which means that 50 per cent of affected patients are homozygous for this mutation (see Chapter 7). Most of the other patients are compound heterozygotes, meaning that they carry two different mutations in the CFTR gene. Besides the ΔF508 mutation, about 1200 different mutations have been found up to 2006, spread all over the gene. A few rare deletions have been described. All types of mutations have been found: nonsense mutations, splice-site mutations, microdeletions and microinsertions causing frameshift, and missense mutations. About 30 mutations have a relative frequency between 0.1 and 5 per cent. Very few *de novo* mutations have been described.

There are geographical and ethnic variations in the way the mutations are distributed. The ΔF508 mutation, found in every population affected by cystic fibrosis, is more frequent in the north than in the south of France, illustrating the gradient found in Europe: in Denmark, it is found in 88 per cent of the mutated chromosomes, but only 50 per cent of the mutated chromosomes in southern European countries. It possibly appeared about 50 000 years ago in an ancestral population of the European populations, and its frequency has been maintained due to a heterozygous selective mechanism. The carrier of the mutations might have been protected against a hydroelectrolytic loss caused by diarrhoeas due to bacterial infections from salmonella type bacteria (this hypothesis has recently been contested). Some mutations are quite frequent in some ethnic groups, like W1282S in Ashkenazi Jews, G551D in Celtic populations, 394delTT in Scandinavian populations. The knowledge of these geographical variations is crucial for the development of molecular genetic tests.

Genotype–phenotype correlations   Genotype–phenotype correlations, meaning the correlations between the mutations and the clinical symptoms, first depend on the mutation types. These can be separated into six classes depending on the protein defect (Figure 2.2).

Globally, class I, II, III and VI mutations have a severe effect, while class IV and V have more moderate effects, even very moderate. The combination between two 'severe' mutations is generally found in typical cystic fibrosis children cases, with a pancreatic insufficiency. The combination between a 'severe' and a 'moderate' mutation is likely to be associated with a moderate form, with a later age of onset, and pancreatic function unaffected. The combination of a 'severe' and a 'very moderate' mutation, or of two 'moderate' mutations, can be found in adults suffering from isolated infertility due to a lack of the vas deferens, or from a mainly respiratory or pancreatic disease expression.

Mutation classification depends on functional studies that can only be performed for a minority of mutations in research laboratories, while new mutations are constantly being found. Among those, the missense mutations can be either deleterious or functional (polymorphisms). Also, most of the studies that have been done use different protocols and different cellular models depending on the laboratory, which

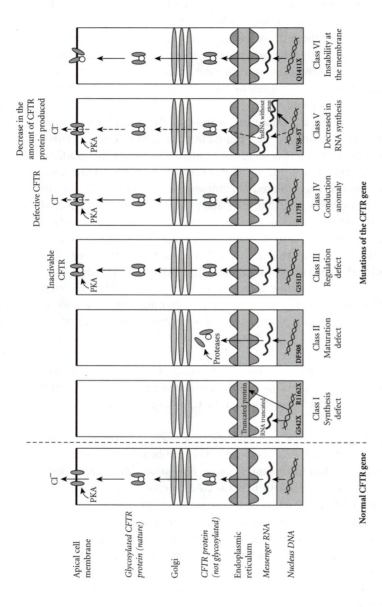

Figure 2.2 The different classes of mutation for the CFTR gene

- Class I groups the mutations affecting the production of the protein; it consists of non-sense mutations, splicing mutations and micro-insertions or micro-deletions causing a frameshift.
- Class II corresponds to the mutations affecting the process of cellular maturation of the protein. The mutation ΔF508 belongs to this class: the loss of the phenylalanine at position 508 modifies the normal folding of the protein, preventing its glycosylation and signalling to the membrane of the epithelial cells. These abnormal proteins are recognized by chaperone proteins from the endoplasmic reticulum, and degraded
- Class III contains the mutations that are located in the ATP binding sites and therefore prevent the opening of the Cl⁻ channels.
- Class IV groups the mutations that alter the conduction of Cl⁻ ions by decreasing the opening timing of the channel or by modifying the selectivity of the channel.
- Class V contains the mutations that are associated with a significant diminution of expression of the normal CFTR protein.
- Class VI is made up of mutations at the end of the gene that are responsible for a large decrease in the stability of the CFTR protein.

makes it difficult to interpret the real effect of a mutation. The biologist has to use several indicators to predict whether the identified defect is a severe or a moderate mutation or a polymorphism.

The variability of the clinical forms can only be partially explained by the heterogeneity of the mutations and genotypes. Even if the correlation between the genotype and the effect on the pancreatic functions is quite high, the respiratory effect is much more variable in patients with the same genotype. This variability could be due to the association of several mutations on the same gene, whose joint effect on the CFTR protein function would be different from their individual effects. Specifically, the presence of 5 thymidine (5T) (instead of 7 or 9), in a splicing accepting site on intron 8 is associated with an aberrant splicing of the mRNA, and therefore with a significant decrease in the level of normal CFTR protein, and can increase the effect of another mutation located on the same allele. For example, the same genotype with the two mutations (ΔF508/R117H) has been found in patients with different phenotypes, one with a moderate form of cystic fibrosis (with a normal pancreatic function) and one with infertility due to the absence of both vasa deferentia. This difference in the phenotypes has been explained by the fact that the R117H-5T allele (variant 5T in *cis* of the mutation R117H) is found in patients having cystic fibrosis, while the allele R117H-7T is more often observed in infertile patients. In addition, other genetic factors (mutations in other genes, called modifiers) and environmental factors can modulate the phenotype.

Molecular diagnosis methods

The cloning of the gene and the identification of the mutations has allowed the setting up of some diagnosis strategies that fit the clinical situation. The techniques used almost all depend on the analysis of PCR amplified DNA fragments of the gene.

Indirect approach   The indirect approach, which depends on the results from a genetic linkage analysis using DNA markers of the CFTR gene (co-transmission between the markers and the mutation, see Chapter 1), still gives a few indications (Figure 2.5) but requires study of the affected child (proband).

Direct approach to diagnosis   The direct approach to diagnosis, which is currently the most frequent, essentially depends on the detection of the anomalies responsible for cystic fibrosis, either in an affected child or in the necessarily heterozygous parents. Therefore, the search for causal mutations can be performed a priori on the parents of a deceased child. When the molecular study of the CFTR gene is requested to give a diagnosis (atypical forms of cystic fibrosis or monosymptomatic adult forms), this is the only approach that can confirm the diagnosis by identifying a mutation on each gene copy.

Among the mutation detection methods, it is possible to distinguish between those based on the search for one or a few specific mutations and those called scanning methods, which can detect any sequence variation within a several hundred base pair DNA sequence (see table linked with Figure 2.6). The scanning methods allow determination of the most frequent mutations responsible for the disease, depending on geographic or ethnic origin, followed by selection of the most suitable tools to perform the molecular diagnosis.

At the moment, most of the laboratories initially use tools designed to detect targeted mutations, mainly kits which can detect quickly between eight and 31 mutations simultaneously (see OLA, Chapter 1, and Figure 2.3), identifying 75 to 90 per cent of the alleles in affected patients (Figure 2.6).

In cases where the molecular study needs to be continued (for example, when only one mutation has been identified in a patient), the scanning tools, still used by

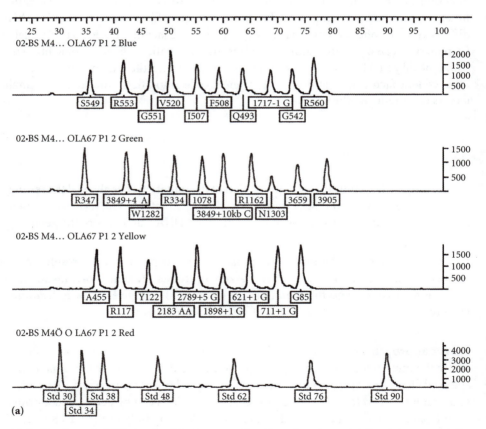

(a) Profile of a control DNA sample without any mutation; the software separates the peaks corresponding to the blue, green, yellow and red fluorescent fragments as four different lines.

Figure 2.3   Detection of CFTR mutations using the OLA technique and chromatography

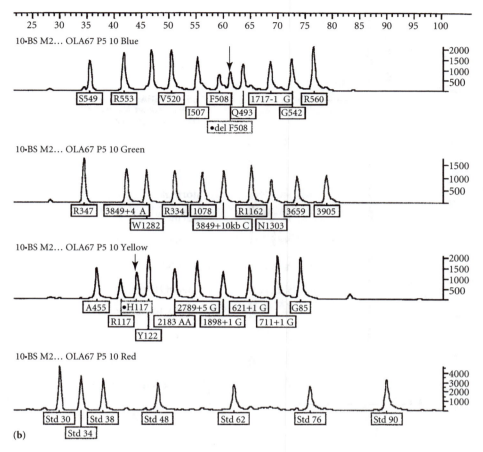

(b) Profile of a DNA sample from a patient affected with a moderate form of cystic fibrosis, composite heterozygote ΔF508/R117H. A diminution of the size of the peak F508 and R117 and the addition (arrows) of the peaks ΔF508 and R117H compared with the standard profile in (a) can be seen.

Figure 2.3 (*Continued*)

a few referenced laboratories, are requested. The 27 exons and the flanking intronic regions are sequentially analysed, allowing the detection of all previously described mutations, as well as the identification of new mutations. These tools take longer to run and are only used where the specific tests fail to give a result. One of the most frequently used methods is denaturing gradient gel electrophoresis (DGGE, see Chapter 1 and Figure 2.4).

Another method, denaturing high-pressure liquid-phase chromatography (dH-PLC), seems very promising because it is semi-automatic. It relies on the differential elution properties of a double-stranded DNA fragment with and without a mutation.

These two methods (DGGE and dHPLC) have a very high sensitivity, mostly for heterozygous mutations. However, even the DGGE method can only detect 95 to

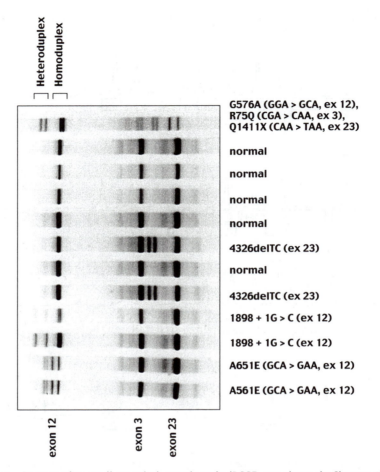

Figure 2.4 Denaturating gradient gel electrophoresis (DGGE, see theory in Chapter 1) profile for three exons of the CFTR gene amplified and analysed simultaneously (multiplex conditions). *Note*: migration from left to right

98 per cent of mutations, and the direct sequencing of the 27 exons cannot identify 100 per cent of the mutations, some still being unknown because they are located within introns that have not been analysed yet. As a consequence, the absence of mutation, or the identification of only one mutation in a patient, cannot rule out a diagnosis for cystic fibrosis or the fact that the CFTR gene is involved in the observed pathology.

The choice by a laboratory of one particular method depends on its sensitivity and reliability, but also on the laboratory history and on the tools already available, as well as on the cost. Each method has its advantages and disadvantages which must be well known by the laboratories (table linked with Figure 2.6), and that can add to the PCR amplification limits. The failure to identify a mutated or a normal allele can be due to: (1) the absence of amplification of this allele because of a deletion of the gene, a mutation or a polymorphism in the PCR primer hybridization sequence or a technical problem; (2) an interference with the mutation research method because of a mutation or a polymorphism in the probe hybridization site or the ligation sites of the two probes. In general, a result suggesting that a patient is homozygous for a mutation can only be confirmed if both parents carry this mutation. This is especially important when the cystic fibrosis diagnosis is not certain, or when a heterozygous diagnosis is requested in the family. If an affected child is a compound heterozygote for the ΔF508 mutation and for a deletion of exon 10, they will appear as if they were homozygous for the ΔF508 mutation. A negative result from the search for ΔF508 in the sister or brother who wants to have children might give false reassurance.

In contrast to point mutations, unknown deletions are not routinely looked for because of their low frequency (globally below 5 per cent) and because of the inability of classical PCR techniques to quantify DNA. A deletion can only be detected in two circumstances:

- in a patient who has a homozygous mutation, because in that case, one or more exons fail to be amplified,

- in a patient heterozygous because of a bad segregation of the intragenic polymorphic markers in the family (Figure 2.5). A deletion can be quickly and efficiently looked for, in homozygotes as well as in heterozygotes, only if its size and limits have been characterized. It is for the detection of previously unknown deletions that the real-time quantitative PCR technique has a considerable interest.

Another promising technique is DNA chip/array technology, based on molecular hybridization, where a patient's DNA would be put together with hundreds (possibly thousands) of specific probes, concentrated in one cm^2 chip. In the case of the CFTR gene, this would allow you to search in one step, either for all of the mutations identified so far – which is not of so much interest knowing that a large number of the mutations are very rare and that some other mutations will soon be identified – or for the most frequent mutations, taking into account ethnic and geographic specificities.

Situations and strategies for the use of molecular tests in genetic counselling for cystic fibrosis

Molecular tests can only be performed after a medical prescription has been written following a genetic consultation; this is a key step in the process. The consultation has to be on an individual basis, and is to inform the patient, or an asymptomatic

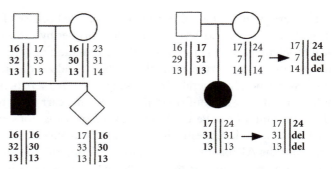

16	17		16	23
32	33		30	31
13	13		13	14

16	16		17	16
32	30		33	30
13	13		13	13

(a) To determine the alleles associated with the disease in order to perform a prenatal diagnosis

16	17		17	24		17	24
29	31		7	7	→	7	del
13	13		14	14		14	del

17	24		17	24
31	31	→	31	del
13	13		13	del

(b) Evidence for a deletion

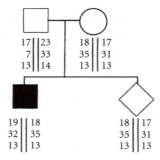

17	23		18	17
7	33		35	31
13	14		13	13

19	18		18	17
32	35		35	31
13	13		13	13

(c) false paternity
(d) sample error

Suspicion of:
(e) maternal contamination
(in case of prenatal diagnosis)

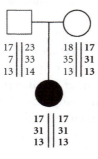

17	23		18	17
7	33		35	31
13	14		13	13

17	17
31	31
13	13

(f) Suspicion of uniparental disomy
→ analysis of microsatellite markers from other chromosomes

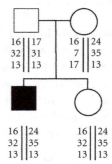

16	17		16	24
32	31		7	35
13	13		17	13

16	24		16	24
32	35		32	35
13	13		13	13

(g) CFTR gene exclusion in the pathology

(a) Indirect diagnosis approach: identify the alleles of markers or the haplotypes (association of different markers on one chromosome) that have been transmitted by the parents to the affected child. In this case, the study is informative and it is possible to offer a prenatal diagnosis. The 'morbid' haplotypes are indicated in bold. The fetus inherited the 'morbid' haplotype from his mother and the normal haplotype from his father; he is not affected.

(b) An 'anomaly' of allele transmission leads to the suspicion of a partial deletion of the gene removing the last two markers: the mother and child appear homozygous for different alleles. The analysis does not allow detecting if an individual is homozygous or hemizygous for one allele.

individual, of the risks of having a child suffering from cystic fibrosis and of the limitations, possible results and consequences of the test. During this consultation, the doctor has to obtain a written agreement from the person who will be tested, or her/his parents if the patient is a minor.

Besides systematic prenatal diagnosis, which is organized in a special way, there are three other types of situation leading to a molecular study of the CFTR gene (Figure 2.6):

- family study (child affected with cystic fibrosis and his parents) in the context of a prenatal diagnosis,

- screening for the relative of an affected patient, or the spouse of a heterozygote,

- help for the diagnosis of atypical or adult forms of cystic fibrosis.

An increasing number of requests concern aid for diagnosis and screening of heterozygotes. Knowing the ethnic and geographical origin of the patient allows the determination of how sensitive a test searching for only a few mutations would be, and the orientation of the study towards a search for the mutations not detected by the test, but occurring frequently in the population the patient is from.

Family studies and prenatal diagnosis For a couple who already have one child affected with cystic fibrosis, the risk to have another affected child is 1/4 for each pregnancy. Before doing the prenatal diagnosis, it is necessary to perform a molecular study of the CFTR gene in the family to detect the alleles linked with the disease using a direct approach or, eventually, an indirect approach. The indirect approach is doable only if the cystic fibrosis diagnosis is certain.

Screening relatives of an affected individual and their spouse The a priori risk for a healthy individual related to an affected patient, to carry a mutation in the CFTR gene, depends on their blood relationship with the patient; 2/3 for the siblings, 1/2 for uncles and aunts, 1/4 for first cousins, etc. The heterozygous frequency for cystic fibrosis in the general population being 1/25, the a priori risk for the couple to have

---

(c) A false paternity is suspected if the paternal haplotype is not found in the child.

(d) A sample error is suspected if a child carries alleles other than those found in the parents.

(e) Contamination of fetal cells by maternal cells can happen when the fetal sampling is performed. If the contamination is significant, the result of the analysis will show the same genotype as the mother.

(f) A uniparental disomy signifies that the two chromosome 7s of the child are inherited from the same parent. A priori, it is the same chromosome 7, carrier of a mutation responsible for cystic fibrosis, present in two copies in the child.

(g) The observation, within the same sibship, of one affected child and one healthy child who share the same parental haplotypes, can exclude the implication of the CFTR gene in the pathology.

Figure 2.5 Indications from a study of microsatellite, RFLP or SNP markers in the CFTR gene

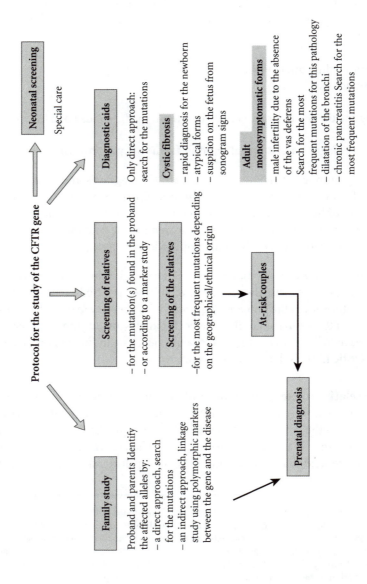

Protocol for the study of the CFTR gene

Neonatal screening

Special care

Diagnostic aids

Only direct approach:
search for the mutations

Cystic fibrosis
– rapid diagnosis for the newborn
– atypical forms
– suspicion on the fetus from
sonogram signs

**Adult
monosymptomatic forms**
– male infertility due to the absence
of the vas deferens
Search for the most
frequent mutations for this pathology
– dilatation of the bronchi
– chronic pancreatitis Search for the
most frequent mutations

Screening of relatives

– for the mutation(s) found in the proband
– or according to a marker study

Screening of the relatives

–for the most frequent mutations depending
on the geographical/ethnical origin

At-risk couples

Prenatal diagnosis

Family study

Proband and parents Identify
the affected alleles by:
– a direct approach, search
for the mutations
– an indirect approach, linkage
study using polymorphic markers
between the gene and the disease

Advantages and disadvantages of methods to detect mutations in the CFTR gene

Targeted detection method	Detected mutations	Advantages	Disadvantages
Heteroduplex analysis (double-stranded molecules with a mismatch) (PCR + control gel)	Mostly ΔF508 and ΔF507. Other micro-insertions/deletions	Fast and simple.	Migrations profile not specific to a particular mutation.
Restriction enzyme digestion (PCR + restriction + control gel)	Depends on the sequence	Fast and simple Identification of a mutation detected by a scanning method (avoids sequencing).	Not specific, especially if the mutation suppresses a restriction site (two neighbouring mutations can suppress the same restriction site: G551D and R553X abolish the same restriction site for HincII).
ASO hybridization (allele specific hybridization) dot blot/reverse-dot blot	8 to 17 mutations (kit)	Fast, automatic Reverse-dot blot: several mutations in one go.	Difficult to set up
ARMS (allele specific amplification)	12 to 20 mutations (kit)	Fast Several mutations in one go.	The choice of the PCR primers is critical. Results based on the lack of PCR products. No distinction between homozygotes and heterozygotes (except for ΔF508 (kit).
OLA (oligonucleotide ligation)	31 mutations (kit)	Fast Several mutations in one go.	Requires use of a DNA sequencer (expensive).
Scanning methods			
DGGE (denaturing gradient gel electrophoresis)	All mutations located in the coding regions and at the intron borders	High sensitivity (>95%).	Difficult to set up Impossible to automate. Risk of missing some mutations when homozygous.
dHPLC (denaturing high-performance liquid chromatography)		Fast, semi-automated.	Bad detections of homozygous mutations. Expensive equipment.
SSCP (single-strand conformation fast polymorphism		Fast and simple.	Sensitivity limited to 80–85%.
Sequencing		100% sensitivity (in theory).	Expensive to use in the first instance.

Figure 2.6 Protocol for the study of the CFTR gene

an affected child is then respectively 1/150 (2/3 × 1/25 × 1/4 = 1/150), 1/200, 1/400. The strategy for performing the study depends on the knowledge of the mutation in the proband, and the urgency of the situation (if there is an ongoing pregnancy or not). The best strategy is to first screen the relative for the anomaly found in the patient. If he does not carry the anomaly, the couple can be reassured. If the relative is heterozygous, the risk for the couple becomes 1/100, whatever the blood relationship, and it is necessary to search the spouse for the most common mutations linked with the disease. As soon as a known deleterious mutation is identified in the spouse, the risk reaches 1/4 and a prenatal diagnosis can be offered. If the screening tests are negative in the spouse, the risk of having an affected child decreases, which allows the parents to be reassured, and avoids the need for a prenatal diagnosis. The absence of the ΔF508 mutation decreases the risk to 1/320, the absence of the eight most frequent mutations (80 per cent of the mutated alleles) decreases it to 1/480, and the absence of the most frequent 31 mutations (85–90 per cent of the mutated alleles) to 1/640–1/960. The minimum risk cannot go below 1/2000, considering that the maximum detection rate is 95 per cent. Knowing the origin of the individuals allows searching specifically for the most frequent mutations to determine precisely the risk of having a child affected with cystic fibrosis. However, it is sometimes difficult to establish a precise risk when the geographical origins are diverse.

Diagnostic aids   Diagnostic aids concern the atypical forms of cystic fibrosis whose diagnosis can be confirmed only if two allelic mutations (inherited from each parent) have been identified, and can eventually allow the proposal of a prenatal diagnosis to the parents. They also concern adult patients who present with infertility due to the absence of the vas deferens, dilatation of the bronchial tubes which cannot be explained, chronic pancreatitis or a naso-sinusal polyposis especially with chronic infection that started during childhood or adolescence. Between 50 and 85 per cent of patients who are infertile due to absence of the vas deferens carry two allelic mutations of the CFTR gene, and about 30 per cent of the patients with a bronchial tubes dilatation or a chronic pancreatitis carry at least one mutation in the CFTR gene. In those situations, there is a double interest in studying the CFTR gene:

• to identify the cause of the disease and, eventually, to recognize an atypical form of cystic fibrosis so as to be able to offer an adapted system of treatment;

• to offer genetic counselling to the patients and their spouses, or their relatives who could be heterozygous for a severe mutation.

In practice, if none of the most frequent mutations is found, the possibility to search for rare mutations is discussed with the clinician. The presence of one of the most frequent mutations in a heterozygous state encourages the search for another mutation to confirm the role of the CFTR gene in the patient pathology.

In the case of male infertility, and when a couple is asking for a medically assisted fertilization, knowing the high proportion of infertile males carrying two mutations and considering the fact that those mutations can take a long time to be identified, it is recommended to first search in the wife for the most frequent mutations. If this search is negative, the couple is reassured. If the wife is heterozygous, it is very important to identify the mutations carried by the man in order to provide appropriate genetic counselling. With an infertile man carrying two mutations, the child will definitely inherit one of the paternal mutations, and will inherit the mother's mutation with a probability 1/2. If the man's genotype combines one severe mutation (like ΔF508) and a very moderate one, like the 5T variant of intron 8, the risk of classic cystic fibrosis for the child is 1/4. If the man is a compound heterozygous for two mutations with a moderate effect, the risk of cystic fibrosis is 1/2 in theory, but it will probably be a moderate or minor form of the disease. However, this is the worst type of prediction, for a moderate form of cystic fibrosis is not necessarily reassuring. A prenatal diagnosis, or pre-implantation diagnosis in the case where a medically assisted fertilization using intra-cytoplasmic sperm injection (ICSI) is considered because of the male infertility, can be perfectly acceptable.

Special mention should be made about couples having no family history of cystic fibrosis and expecting a child suspected to have the disease because of echogenic bowel seen on the sonogram, suggesting an intestinal blockage (bowel obstruction). If the pregnancy is at less than 20 weeks, measuring the intestinal enzymes in the amniotic fluid is helpful. A normal dosage before this term can rule out cystic fibrosis; extremely low values reflect a digestive obstruction that can be due to cystic fibrosis. However, most of the time, the pregnancy is more advanced, and measuring the intestinal enzymes in the amniotic fluid is not interpretable. A search for an anomaly in the CFTR gene, in parallel with the fetal karyotype and a specific viral serology and parasite serology to find causes other than cystic fibrosis, is worth doing only if the couple is well informed and is asking for a medical pregnancy interruption in case the diagnosis is positive (3 per cent of echogenic bowels are due to cystic fibrosis); this request largely depends on the term of the pregnancy. Only the identification of two mutations in the fetus, inherited from both parents, gives a diagnosis for cystic fibrosis. In practice, the search for the mutations is first made in the parents. If the ΔF508 mutation is the only one looked for and is not detected, the risk of the fetus being affected with cystic fibrosis reaches about 1/350; and about 1/1400 if none of the 31 frequent mutations is detected. The discovery of a mutation in only one of the parents requires searching for this mutation in the fetus: if the fetus did not inherit the mutation, the diagnosis of cystic fibrosis is ruled out. However, if the fetus carries the mutation, the risk for him to be carrying another mutation undetected in the first test made in the parents is significant (11–13 per cent after the test for the 31 mutations), justifying a search for a second mutation.

Conclusions   Since the CFTR gene was discovered in 1989, important progress has been made in developing tools to study this gene. These permit characterization of a considerable spectrum of mutations and to determine their geographic and ethnic distribution and, from there, to set up efficient strategies for molecular diagnosis and genetic counselling, and also to show that mutations in the CFTR gene are responsible for very heterogeneous clinical forms. Despite this progress, the demonstration of the deleterious or neutral character of many mutations is impossible to do routinely, which sometimes makes genetic counselling difficult. Investigation into the correlation between genotypes and phenotypes shows that a large number of factors – genetic and environmental – modulate the expression of the CFTR gene mutations (see page 96 of French edition).

2.2.2 Haemochromatosis

Haemochromatosis is the most frequent hereditary disorder, affecting between three and four individuals out of 1000 in Caucasian populations. The discovery of the HFE gene in 1996 and the setting up of a simple genetic test have considerably modified the diagnosis for this disease.

Clinical and biological aspects of haemochromatosis

The term haemochromatosis was first used by Von Recklinghausen in 1889 to describe a generalized iron overload in the organism, leading to many visceral and metabolic complications. The facts that the disease was hereditary and was transmitted recessively were only shown in 1975.

   Haemochromatosis, characterized by an increase in intestinal iron absorption, is a late onset disease. The first phase is latent, then some biological non-symptomatic anomalies appear such as an increase in the serum iron level, in the transferrin saturation coefficient and finally, in the serum ferritin level. Clinical signs tend to appear in men between the ages of 30 and 40 and slightly later in women due to menstruation and pregnancies. When fully expressed, the clinical picture of the disease associates many symptoms in correlation with a general excess of iron:

• skin and body signs dominated by a melanodermy, typically metallic grey, possibly with ichthyosis, nails abnormalities and an abnormal hair smoothness;

• hepatopathy with cirrhosis leading to a hepatocarcinoma in one third of the cases;

• frequent arthropathy – the most characteristic expression being chronic arthritis of the second and third metacarpal-phalange joints;

• type I diabetes (insulin dependant diabetes);

- other endocrine disorders are rare, mostly affecting the gonads: early menopause in women, impotency and testicular atrophy in men;

- rare effects on the heart, with dilated cardiomyopathy.

This typical clinical picture is rarely observed. Currently, the diagnosis of haemochromatosis is more often suggested by less characteristic chronic signs such as asthenia, arthralgia and moderate increase in transaminase levels.

For all cases, the treatment consists of regular venesections whose spacing and number is determined by the initial medical examination and, in particular, the serum ferritin level. The benefit of the depletive treatment depends on how early the disease was diagnosed. At an advanced stage, only the cardiomyopathy is sensitive to the treatment while most of the other complications (cirrhosis, diabetes, arthropathies and endocrinopathies) do not improve much. In contrast, when the diagnosis is made before the stage of tissue lesions, the life expectancy of the patients is close to that of the general population.

Genetic determinism of haemochromatosis

One gene responsible for haemochromatosis, named HFE1, was localized to 6p21.3 in 1975 on the basis of a strong linkage to the HLA-A3 antigen, and was cloned in 1996 (Feder *et al.*, 1996). The definitive implication of this gene in haemochromatosis was shown by the description of mice which had the HFE1 gene knocked-out (inactivated) and an iron overload in their liver the same as the patients with haemochromatosis.

Genetic heterogeneity: one disease/several genes   The discovery of the gene HFE1 leads to the establishment of a new classification of the primary iron overloading, from which three are well identified:

- haemochromatosis HFE1 is in first place because more than 90 per cent of the patients showing clinical signs of haemochromatosis carry two mutated copies of HFE1;

- the childhood form of haemochromatosis (HFE2), a special form of haemochromatosis with an earlier onset, is characterized by major heart and endocrine complications – two genes responsible for HFE2 have been localized and cloned: HJV on the long arm of the chromosome 1 (Roetto *et al.*, 1999; Papanikolau *et al.*, 2004) and HAMP on the long arm of chromosome 19 (Roetto *et al.*, 2003);

- haemochromatosis (HFE3), described in patients from Sicily with a typical clinical picture: these patients are homozygous for the Y250X mutation in the TfR2 gene (transferrin receptor 2) localized on 7q22;

- dominant haemochromatosis (HFE4) involves the SLC11A3 gene (also called Ireg 1 or ferroportin);

- two forms for which the genes have not been localized or identified: first, neonatal haemochromatosis, characterized by a major hepatic iron overload leading to death during the neonatal period; the second form is iron overload in African populations, considered up until recently as having an alimentary origin;

- multifactorial forms associating several special genotypes for the HFE1 gene and environmental factors.

Haemochromatosis involving the gene HFE1   The HFE1 gene extends over a 12 kb region, contains seven exons and codes for a 343 amino acid protein with high similarities to the major histocompatibility complex class I HLA molecule (Fergelot *et al.*, 1998) (Figure 2.7(a)).

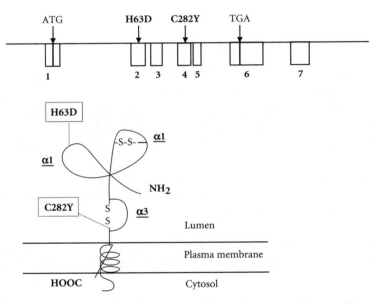

The translation initiation codons (ATG) and stop codons (TGA) are found respectively at positions +1248 and +8033 based on the first known base of the gene.
Correspondence between the genomic sequence and the protein domains
exon 1 (296 bp) : 5′ UTR and peptide signal
exon 2 (263 bp) : extracellular domain α1
exon 3 (275 bp) : extracellular domain α2
exon 4 (275 bp) : extracellular domain α3
exon 5 (113 bp) : transmembrane domain
exon 6 (1055 bp) : cytoplasmic tail and 3′UTR
exon 7 (431 bp) : 3′UTR

Figure 2.7   Gene and protein organization for SFE, positions of the C282Y and H63D mutations

- The main mutation C282Y, located on exon 4, leads to the replacement of cysteine 282 by a tyrosine. This mutation is found in the homozygous state in more than 90 per cent of patients with haemochromatosis in north-west Europe (Jouanole *et al.*, 1996). This frequency decreases following a north-west to south-east gradient to reach 40 per cent in the south of Italy. It is specific to European populations, or populations with a European origin, in which the allelic frequency in the general population reaches 7 per cent. It is not found in Africa or in Asia.

- Private mutations, that are very rare: a mutation in the splicing site (IVS3 + G/T) causing the elimination of exon 3; two missense mutations in exon 3 (E168X and W169X) can be the result of a founder effect in the north of Italy.

- Two main polymorphic variants:

  (a) H63D, in exon 2, leads to the replacement of a histidine with an aspartate in the protein sequence. This variant is mutually exclusive with the C282Y mutation (the two are never present on the same chromosome) and is found with a frequency of 16 per cent in the general population. It should be considered as a single nucleotide polymorphism (SNP) that can modulate the iron overload in association with mutation C282Y (Beutler, 1997).

  (b) S65C, also in exon 2, leads to the replacement of a serine with a cysteine in position 65 of the protein sequence. Its allelic frequency is 3.2 per cent in the general population (Douabin *et al.*, 1999). The implication of this variant in iron overload is still controversial.

The HFE1 protein

Although the HFE gene was described in 1996, the role of this protein in mechanisms of intestinal absorption and mechanism of the tissue overload are still not fully understood.

- The HFE1 protein belongs to the family of class I HLA molecules and therefore consists of three extracellular domains (α1, α2, α3), one transmembrane domain and a short intracytoplasmic tail. Crystallization studies show that they cannot bind to endogenous peptides, in contrast to the classic class I HLA molecules. However, its transport to the cell surface involves a covalent bond with b2 microglobulin; this bond is broken by the C282Y mutation (Figure 2.7(b)).

- Immunohistochemical studies reveal the presence of the HFE protein in the epithelial cells of the digestive tract: at the level of the plasma membrane of the non-polarized cells of the oesophagus and at the basolateral membrane of the polarized epithelial cells of the stomach and the colon. In the intestinal crypts, the protein is mostly perinuclear. It does not seem to be present in the hepatocytes, or in the mononuclear phagocyte system cells.

- The HFE protein can form a complex with the transferrin receptor (TfR), decreasing the affinity of this receptor for its ligand and, therefore, the entry of the iron into the cell. This HFE/TfR interaction is not sufficient to explain in a coherent fashion the role of HFE in the regulation of intestinal iron absorption that happens at the level of the duodenum and in the cryptal enterocytes of the jejunum. The hypothesis is made that in the crypt cells, HFE could play a sensor role, programming the iron absorption at the level of the enterocytes depending on the iron content of the organism.

Molecular diagnosis of the HFE1 gene mutation identified

The gold standard method is PCR-restriction (see Chapter 1), easy to set up in any molecular genetics laboratory.

Detection of the C282Y mutation   Localized in HFE gene exon 4, it creates a restriction site for the *RsaI* restriction enzyme. An exon4 sequence covering the mutation site is amplified by PCR using primers located on both sides of the region of interest. The amplified fragment corresponding to the normal allele (387bp) has a single restriction site for *RsaI* at the position 247; the mutated allele has an extra site for *RsaI* (Figure 2.8). After the PCR, the amplified fragment is incubated with *RsaI* and the analysis of the restriction profile after electrophoresis gives the patient genotype.

Search for the H63D variant   The H63D mutation, localized in the HFE gene exon2, removes one restriction site for the *MboI* enzyme. A genomic sequence of 208bp covering the mutation site is amplified by PCR and digested by the *MboI* enzyme. A normal allele would give two fragments of 138 and 70 bp while the mutated allele will give the same size as the amplified fragment, that is to say 208bp (Figure 2.9).

Diagnostic strategies

They only concern the diagnosis for HFE1 Haemochromatosis, defined as follows.

The positive diagnosis   In most cases, the diagnosis is suggested by clinical signs more or less evident or by biological signs like moderate hypertransaminasaemia or a perturbation of the iron level.

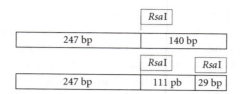

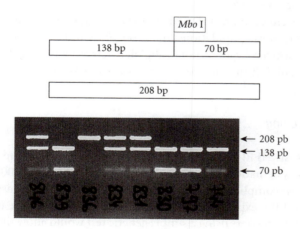

Top: amplicon from the normal allele. bottom: amplicon from the C282Y allele.
Control electrophoresis of the amplicons after restriction with the enzyme *Rsa*I (3 per cent agarose).
zone a : 247 bp bands; zone b : 140 and 111 bp bands.

Figure 2.8   Detection of the C282Y mutation

The second step consists of an evaluation of the transferrin saturation coefficient that constitutes the most specific biological marker for haemochromatosis. A normal level (<30–40 per cent) indicates a negative diagnosis in the absence of bleeding or of inflammatory syndrome.

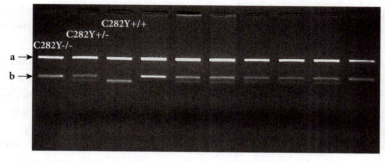

Top: amplicon from the normal allele. Bottom: amplicon from the H63D allele.
Control electrophoresis of the amplicons after restriction with the enzyme *Mbo*I (3 per cent agarose).

Figure 2.9   Detection of the variant H63D

The third step requires a genetic test with the search for the C282Y mutation. If the patient is C282Y homozygous, haemochromatosis can be affirmed. The intensity of the iron overload, in general well correlated with the ferritinaemia, still has to be evaluated. When ferritinaemia is below 1000µg/ml, in the absence of hepatomegalia and of hepatic cytolysis, the risk of cirrhosis or of fibrosis is null and the hepatic biopsy puncture gives no indications of problems. However, when the ferritinaemia is above 1000 µg/ml, the hepatic biopsy puncture must be performed for prognostic purposes because a fibrosis is observed in half of the cases (Brissot *et al.*, 2000).

In a few rare cases, faced with an apparently classic case of haemochromatosis not associated with C282Y homozygosity, mutations can be looked for by a complete sequencing of the HFE1 gene, or even of TfR2. Finally, when signs indicate childhood haemochromatosis, the sequencing of the HJR or HAMP genes (localized on 1q and 19q respectively) could be performed to identify mutations.

Comment: Despite the presence of clinical or biological signs suggesting haemochromatosis, the genetic test can be negative (C282Y−/−) or can reveal the presence of the C282Y mutation in a heterozygous state (C282Y +/−). In this case and on the request of the practitioner, the H63D variant can be looked for. The genotype C282Y/H63D can favour iron overload, in conjunction with other genetic or iatrogenic factors, but does not allow genetic counselling for the family

Genetic counselling   This relies on the search for the C282Y mutation in the family of a homozygous proband, and concerns primarily the proband's brothers and sisters. Any C282Y homozygous patient will have to perform an exhaustive clinical and biological examination and, eventually, a regular biological surveillance if no expression is observed.

The proband's children are at least heterozygous, but considering the frequency of the mutation in some populations, the probability of having a union between a homozygote and a heterozygote is not negligible, thereby increasing the risk of having an homozygous child. It therefore seems necessary to test the spouse first.

Systematic screening

Haemochromatosis HFEI has all the characteristics normally considered for a disease suitable for systematic screening: frequent (three to four per 1000), potentially serious, and whose complications can easily be avoided through an easy and low-cost treatment. Even if the expressivity of the C282Y homozygous state is variable, the identification of the carriers using a simple genetic test would allow the start of treatment before the appearance of the first biological signs and avoid the complications of the disease.

2.2.3   Thalassaemias and drepanocytosis

Introduction

Haemoglobinopathies are among the most frequent monogenic diseases in humans with more than 5 per cent of the world population possibly carrying a clinically detectable haemoglobin (Hb) anomaly. This particularly large distribution is due to the selective advantage given by some of these diseases against malaria that was (and still is in some regions) very important. Due to population migration, these anomalies are also very frequent in most large urban centres where they create problems for the genetic diagnosis of these diseases.

Most of these diseases are inherited as recessive traits and lead to chronic anaemias with variable severity that can lead to extremely serious disorders with a complete dependency on blood transfusions. Treatments are being evaluated like the hydroxyurea induction of fetal haemoglobin that can compensate for the absence or an abnormality of the β-globin chain. However, the only therapies against the serious forms are blood transfusions for life, associated with the prevention of the side effects of such a treatment (iron overload, infections, etc.), or a bone marrow transplant.

The clinical severity of the diseases, and the absence of any real therapy, often justifies performing prenatal diagnosis. In some countries (Sardinia, Cyprus) where the prevalence is very high, the treatment cost has lead to an active policy of genetic screening of carriers and the setting up of genetic counselling with the possibility of prenatal diagnosis to reduce greatly the frequency of homozygous patients.

Erythrocyte biology

The way to make a diagnosis for a patient carrying or thought to carry a 'haemoglobinopathy' is linked to the biology of the erythrocyte, the last stage of differentiation of the erythroid lineage that presents two main characteristics at the end of its differentiation.

- It has lost its nucleus and all of its organelles (mitochondria. . .) and 'survives' thanks to the messenger RNA accumulated during the last steps of its development. Because of its energetic metabolism it uses anaerobic glycolysis exclusively; anomalies in this pathway (pyruvate kinase deficiency for example) are themselves causes for chronic anaemia.

- About 30 per cent of its weight is constituted of haemoglobin (Hb), one heterotetramer composed of two alpha globins (α-globin) and two non-alpha globins (β-globin). Two gene families, deriving from the tandem duplication of the same ancestral gene, code for these two different types of chains (in 16p13.3 for the α-globin genes and in 11p15.3 for the non-alpha globin genes).

Their activation leads to the production of equilibrated quantities of the two types of proteins. There is an embryonic α-globin gene (zeta) and two α-globin genes expressed during the fetal stage and adult stage (α2 and α1). There is an embryonic β-like globin gene (epsilon), two fetal genes expressed after epsilon and two adult genes, δ and mainly β. Even though the α and β chains have very different primary structures (141 and 146 amino acids respectively, and 44 per cent homology overall), their tertiary structures are very similar.

Each α chain associates tightly with a β chain and then two α–β dimers associate less tightly to produce the tetramer. Hb's affinity for oxygen (one molecule at each haem, located in a pocket of each globin) is in relation with a modification of its quaternary structure, the low affinity form (deoxyhaemoglobin) being more compact ('tense' or T form) than the high affinity form ('relaxed' or R form). This variation of affinity follows a sigmoid curve influenced by several elements. The first of these elements is the fixation of the first oxygen molecule (allosteric interaction) that permits the transition from the T form to the R form and, therefore, the rapid fixation of the other three oxygen molecules. Among the other elements, a product from glycolysis, 2.3 diphosphoglycerate (2.3 DPG), intercalates between the two α–β dimers and stabilizes the T form, deoxyhaemoglobin, with low affinity for oxygen.

As a consequence, a haemoglobinopathy can result either from a qualitative defect of the haemoglobin (a chain with an altered capacity to bind the haem or that has its interaction properties with the other chains modified), or from an insufficient production of one of the two chains, α 'non-α', giving respectively the α–thalassaemia and β–thalassaemias. In the same way, an anomaly in the glycolysis pathway (increasing the 2.3 DPG concentration) or in the haem production (iron deficiency) will have an effect on the consequences of a haemoglobinopathy.

Phenotypic diagnosis of haemoglobinopathies

The phenotypic diagnosis of haemoglobinopathies is a useful step for their genetic diagnosis. Indeed, the study of the erythrocyte phenotype often allows determination of the gene involved and of the type of anomaly in a way that orientates the molecular test toward the locus implicated. The phenotypic diagnosis also allows the detection of associations with mutations affecting different loci, associations that happen frequently in some population groups (HbS and α–thalassaemia).

The blood count   The blood count allows the estimation of the haemoglobin (Hb) quantity, the number of red blood cells (RBC), the mean corpuscular volume (MCV), the red blood cell distribution width (RDW) and the mean corpuscular haemoglobin concentration (MCHM). Anaemia is defined by a haemoglobin concentration below

120g/l in men and 110 g/l in women. The reduction of this value is proportional to the number of altered genes.

The mean cell volume (MCV) is an important value allowing for the appreciation of the intramedullary dyserythropoiesis that characterizes certain thalassaemias. This parameter also allows the determination of the role of an iron deficiency in chronic anaemia, if it has decreased significantly. An increase in the red blood cell distribution width (increase in RDW) is also an element in favour of an iron deficiency as the origin of the anaemia.

Haemoglobin study The study of haemoglobin allows the orientation of the genetic exploration by showing an abnormal ratio and by estimating precisely the quantities of fetal Hb ($\alpha2\gamma2$) and HbA2 ($\alpha2\delta2$), two minor fractions frequently increased in haemoglobinopathies. The HbA2 value is very important for the thalassaemia diagnosis (see later) because it is almost always increased in the case of severe β-thalassaemia (>3 per cent of the total Hb), while it stays at normal levels in α-thalassaemia. Hb F ($\alpha2\gamma2$) will be the only haemoglobin in severe β-thalassaemia with HbA2, and an element in the diagnosis of hereditary persistence of fetal Hb (HPFH) otherwise. The existence of HPFH is a prognostic element for haemoglobinopathies, because in a certain number of cases it compensates for the consequences of the Hb anomaly. The existence of Hb H ($\beta4$) or Hb Barts' ($\gamma4$) is a diagnostic element for severe α-thalassaemia with three out of four α-genes deleted.

Even though it is a very important step, the haemoglobin study can be of no benefit unless it uses different techniques whose results, when compared, allow the detection of different variants. For example, electrophoresis at alkaline pH does not allow the differentiation of the A2, E, C and O haemoglobins or the S, D and G haemoglobins. The isoelectric focalization of the Hb allows a better differentiation of the different abnormal haemoglobins but is only performed in very few laboratories. The Itano test allows a formal identification of HbS, taking advantage of a particularity in its solubility, which allows differentiation with another haemoglobin variant which co-migrates with it during a simple electrophoresis (for example HbD).

Molecular diagnosis of genetic anomalies of haemoglobin

The small size of the implicated genes (three exons and less than 2000 base pairs for the genomic sequence of the α-type and β-type genes) is beneficial. However, 800 qualitative mutations have been counted, affecting the coding sequence of the β–globin and α–globin genes, and leading to more or less pathogenic haemoglobin variants. Two hundred thalassaemic point mutations (quantitative) have also been counted, affecting one of the expression steps of the mutated gene (transcription, splicing, polyadenylation, stability and translation of the messenger RNA) or partial or complete deletion of the gene. Also, a few gene fusions due to unequal cross-overs are seen.

Depending on the clinical, cellular and biochemical diagnosis, the molecular analysis will be oriented toward the search for mutations of a particular type in a particular gene, and for this various methods can be used.

Methods for the search of a known mutation

- Restriction site analysis: PCR followed by enzymatic digestion. If the mutation creates or removes a site, or if a site is created by a specific primer conditional on the presence of the searched for mutation (see Section 1.9.8).

- Allele-specific oligonucleotide hybridization (ASOH, see Section 1.9.8): PCR followed by hybridization of oligonucleotides specific for the mutations fixed to a membrane support (reverse-dot blot) or mixed with PCR product immobilized on a support (dot blot).

- Allele specific amplification: PCR using primers whose terminal 3' nucleotide is specific for the nucleotide variant (ARMS, see Section 1.9.7).

Methods for the search of unknown mutations

- PCR followed by a migration technique DGGE or SSCP (see Section 1.10).

- Direct sequencing of the three exons of the gene, and of the intronic regions involved in splicing, of the 5' promoter region and of the 3' region carrying the polyadenylation signal.

Indirect method   The indirect method of family analysis (see Section 1.9.8, Figure 1.21) is useful when previous methods failed or are not applicable, but requires DNA from the proband. It is easily applicable in the case of mutations in the β-globin gene because several polymorphic restriction sites define restriction haplotypes in the region of these genes. In a family carrying a mutation in the β-chains, if there is any informativity, it is possible to use these restriction polymorphisms to follow the chromosome carrying the mutation and to determine 'carrier' or 'non-carrier' status of a patient (at the risk of a recombination). Four polymorphic sites are located within the β-globin and allow the use of this approach with a recombination risk that is almost zero. The modification of the stability of the amplified fragment associated with intragenic polymorphic sites is especially well exploited by the DGGE method.

Molecular diagnosis of the main mutations of the β–gene

More than 434 variants of the β-chain have been characterized. Most of these variants are silent in their clinical aspect and rare. Only three frequent β-globin variants, responsible for severe pathologies and that are therefore often looked for, will be discussed here.

Table 2.1 Evolution of HbS and HbA2 rates in heterozygotes depending on the number of α–globin genes

Number of α genes	%HbS	%HbA2
5 genes : $\alpha\alpha\alpha/\alpha\alpha$	45 to 60%	<3%
4 genes : $\alpha\alpha/\alpha\alpha$	40 %	3.8%
3 genes : $-\alpha/\alpha\alpha$	35%	4%
2 genes : $-\alpha/-\alpha$ or $-\alpha\alpha$	30%	4.5%
1 gene : $--/-\alpha$	20%	4.5%

Sickle cell anaemia mutation   This consists of a missense mutation at codon 6, GAG→GTG, leading to a Glu→Val substitution. The presence of this mutation leads to the formation of haemoglobin S (HbS), completely in homozygotes, and at a minor concentration in heterozygous healthy carriers. When the HbS concentration is above 50 per cent of the total Hb, its deoxygenated form has the property to polymerize, which deforms the red blood cells and makes them rigid.

In the homozygous state, HbS is responsible for sickle cell anaemia, a severe disease that consists of the destruction of the red blood cells in the spleen and vascular microthrombosis responsible for multiple side effects (articular, central nervous system, renal, etc.). The severity of this disease can justify a prenatal diagnosis.

Due to a lower affinity of the βS-chain for the α-chains compared with the βA-chain, the HbS represents about 35 to 40 per cent of the total Hb in heterozygotes. An associated α-thalassaemia leads to a decrease of the HbS level in the red blood cell. This peculiarity allows the identification of the α-thalassaemia associated with HbS, an association that is especially frequent in some African populations. Table 2.1 gives the approximate HbS value depending on the number of α-genes present.

The genetic diagnosis of the βS mutation uses two of the approach types mentioned above. It can be shown by allele specific hybridization (Figure 2.10(b)), or by PCR-restriction (Figure 2.10(c)). Indeed, the mutation GAG→GTG modifies a restriction site recognized by several restriction enzymes (MstII, Bsu36I, BseRI) that can be used to identify the mutation on a PCR fragment surrounding exon 1 of the β–globin gene. Due to the presence, in some populations, of the sickle cell anemia mutation together with another mutation affecting the same codon (β6 GAG→AAG), and responsible for HbC haemoglobin, the enzymes Bsu36I and BseRI are preferred because they allow the differentiation of the two mutations.

Note 1 It is the deoxygenated form of HbS that is susceptible to polymerization. A deficiency in pyruvate kinase that leads to an increase in 2–3 DPG will then favour its polymerization and induce a clinical phenotype of sickle cell anemia in the βA/βS heterozygote.

Note 2 In the heterozygous state, the βS chain is responsible for a pathologic manifestation only in extreme physiological situations favouring its polymerization (dehydration, hypoxia, etc.). However, some others mutations of the

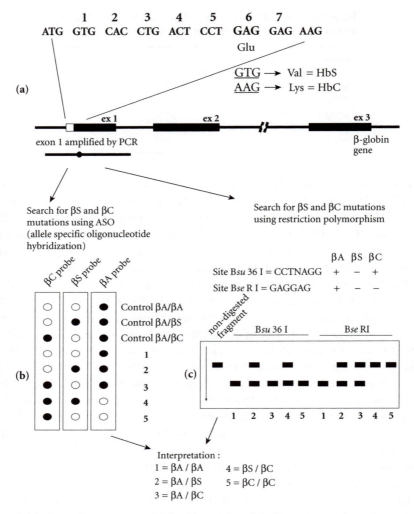

At the top of the figure, β–gene structure, nucleotide sequence of the first seven exons, mutations sequences of the exon 6 leading to the presence of HbS or HbC.

Figure 2.10   Genetic diagnosis of sickle cell anaemia syndromes

β-globin chains can give a pathogenic effect to the presence of a single dose of βS, creating the sickle cell anemia syndrome. It is very important to know of these mutations because, when unknown, they can lead towards a false prenatal diagnosis due to the fact that the only anomalies looked for in these kinds of studies are the ones identified in the parents as being potentially responsible for a severe pathology in the child to be born.

HbC mutation   The HbC mutation also affects codon 6 but leads to a Glu to Lys substitution (GAG→AAG). It gives a haemoglobin variant named HbC (when both β–chains are mutated). When in the homozygous state, this variant (frequently found in Africa) is responsible for a haemolytic anaemia less severe than sickle cell anaemia (HbS), and without any microthrombotic type symptoms. When in the heterozygous state, the abnormal chain represents about 45 per cent of the total Hb and its concentration evolves like those of HbS in the case of combination with an α–thalassaemia.

The molecular diagnosis of the HbC mutation relies on the same techniques as for the βS-chain (Figure 2.10).

HbE mutation   The HbE mutation affects codon 26 (GAG→AAG) and leads to a Glu→Lys substitution. It is responsible for the presence of a haemoglobin variant named HbE, especially frequent in south-east Asia, the mutation having appeared in Cambodia. The βE-chain has a very similar function to the normal chain and its pathogenicity results from the formation of an abnormal splicing site responsible for 10 per cent of messenger RNA being abnormal. The phenotype of this variant is then a phenotype [β-thalassaemic] with a moderate anaemia (see below). Its molecular diagnosis relies on the same techniques as for βS-chain. The mutation modifies a restriction site for the enzyme MnlI.

Compound heterozygotes   In the βS/βC compound heterozygote, a severe pathology, close to sickle cell anemia, results in microthrombosis. It can justify a prenatal diagnosis. Other β-chain variants are susceptible to be pathogenic when combined with the βS-chain. The most frequently found are the mutations Hb O-arab : β 121 GAA→GGA (Glu→Lys) and HbD-punjab : β 121 GAA→CAA (Glu→Gln). S/β thal. compound heterozygotes have a phenotype whose severity depends on the type of the β-thalassaemia mutation (see below). The phenotype is the same as a βS/βS homozygote if the β-thalassaemia mutation is β⁰(with micro-cytic red blood cells and 80 to 95 per cent HbS, 2 to 20 per cent HbF and a rate of HbA2 above 3 per cent) or a less severe sickle cell anaemia phenotype depending on the level of HbS if the thalassaemia mutation is of β+ type (55 to 75 per cent of HbA, 10 to 30 per cent HbA and 3 to 6 per cent HbA2) The severity of these combinations make their detection specially important in the context of prenatal diagnosis.

The thalassaemia mutations   The thalassaemia mutations are mutations leading to a deficiency, or even a loss, of one of the two types of chains, α or β. The resulting syndrome is very frequent in the eastern Mediterranean, and for this reason it was named β-thalassaemia (*thalassa* is the word for sea, in Greek).

In the case of a β-chain deficiency, the extra α-chains do not have the possibility to associate with each other but precipitate, right at the beginning of globin chain synthesis (medullary erythroblast). This precipitate is at the origin of many anomalies that lead to the erythroblast destruction, making erythropoiesis inefficient, which then leads to anaemia due to a deficiency in medullary production (central anaemia).

All types of mutations able to affect gene expression quantitatively have been isolated in β-thalassaemias. Creation of a stop codon, frameshift, appearance or disappearance of a splicing site, defect in promoters, defect in maturation sites of the messenger RNA (capping site, splicing site, polyadenylation site), translation signals (ATG initiator, missense mutation, modification of the stop site). Besides the point mutations, there are a large number of deletions (partial or total).

The mutations are divided into 'β+thalassaemia' and 'β⁰ thalassaemia' depending on the presence of a residual synthesis of β⁻ chain; more than 200 mutations have been described.

The difficulty for the diagnosis of such a large number of β⁻ thalassaemia mutations is reduced by the fact that about 30 of them are responsible for most of the thalassaemias, and by the fact that some of these mutations are grouped depending on the geographical origin of the patient. Table 2.2 gives the main mutations depending on the geography.

Molecular diagnosis of the β-thalassaemia mutations will use a search approach for known mutations if the patient belongs to a geographical zone in which it is possible to have a substantial coverage (over 85–90 per cent) with only a few mutations, as is the case for β-thalassaemia in south-east Asia.

A certain number of screening tests dedicated to a 'region mutations' group are available and are based on techniques using 'multiplex PCR' that is hybridized on

Table 2.2   Main β-thalassaemia mutations depending on the geographical origin

North Africa/Maghreb	Type	Middle East	Type
c 6, −A	β°	IVS 1 nt 110 G > A	β°
Ivs 1 nt 1 G > A	β°	IVS 1 nt 5 G > A	β+
IVS 2 nt 1 G > A	β°	IVS 1 nt 6 T > C	β+
IVS 2 nt 745 C > G	β+	c 39 C > T	β°
c 39 C > T	β°	c 8 del AA	β°
IVS 1 nt 6 T > C	β+		
IVS 1 nt 110 G > A	β°		
India		**Mid and Southern Africa**	
IVS 1 nt 5 G > C	β+	−29 A > G	β+
Del 619 bp	β°	−88 C > T	β+
c 8-9 + G	β°	c 24 T > A	β+
IVS 1 nt 1 G > T	β°	Poly A, T > C	β+
c 41-42 del CTTT	β°		
South-east Asia		**Southern China**	
Hb E (c 26 G > A)	β+	c 41-42 del CTTT	β°
IVS 1 nt 5 G > C	β+	IVS 2 nt 654 C > T	β°
c 41-42 del CTTT	β°	c 17 A > T	β°
c 17 A > T	β°	c 71-72 +A	β°
IVS 2 nt 654 C > T	β°	IVS 1 nt 5 G > C	β+
c 19 A > G	β°	Hb E (c 26 G > A)	β+
c 71-72 + A	β°		
−28 A > G	β+		
c 35 C > A	β+		
−86 C > G	β°		
c 14-15 + G	β°		

oligonucleotide probes corresponding to the mutants present in the population, immobilized on a membrane (reverse-dot blot).

In the case that the number of suspected mutations is too large, it is easier to use simple scanning methods (DGGE or sequencing of the gene). The use of DGGE in this case will allow the association of the mutation with a restriction haplotype internal to the β-globin gene (the framework β-globin) and if the family is informative, it will also allow the realization of the indirect diagnosis for the presence or absence of a mutation with an almost zero risk of recombination. In addition, there is strong linkage between some restriction haplotypes internal to the β-globin gene and some β-thalassaemia mutations, this can help direct the search for mutations, knowing the geographical origin and the β-globin haplotype of the patient.

If point mutations are the most frequent causes of β-thalassaemia, there are some cases in which the cause is a deletion, which can be small or large, removing a β-globin gene (deletional β^0-thalassaemia) or the β– and δ-globin genes ($\delta\beta^0$-thalassaemia) or the γ, δ and β genes ($\gamma\delta\beta^0$-thalassaemia). The diagnosis of these anomalies can be performed by PCR if the boundaries of the deletion are constant enough to define pairs of primers surrounding the break points, or by Southern blot.

Molecular diagnosis of the main mutations of the α-gene

One particularity of the α–globin gene family is the presence of a tandem of genes active at the same time in adults (the α2–gene is twice as active as the α1–gene). Therefore, four genes participate normally to the production of the α–chains. **Qualitative anomalies** 436 α variants have been described, but most of the time they have almost no effect on the phenotype. Indeed, the presence of four active genes make the presence of an abnormal chain very minor and therefore without clinical consequences. The only α variants clinically effective are the unstable ones able to destabilize the tetramer.

Quantitative anomalies   In contrast to the β^- thalassaemias, the α-thalassaemias do not lead to inefficient medullary erythropoiesis. Indeed, the excess β–chains can associate to form 'homotetramers', either of γ–chains (γ4 = Hb Bart's) immediately after birth, or of β–chains (β4 = Hb H) after a few months of life. These two abnormal Hb are not functional because they have an oxygen affinity that is too strong, but they do not induce important defects in the cell. Progressively, these chains will form a precipitate (Heinz's body) that will cause the destruction of the cells in the spleen (peripheral haemolysis). Depending on the number of genes that stay active, one can distinguish:

- the α–thalassaemia trait with three functional genes almost without clinical sign;

- minor α–thalassaemia when two α–genes are functional (microcytosis with Hb A2 normal);

- haemoglobin H disease with only one functional α–gene (severe microcytic anemia with 3 to 30 per cent of Hb Bart's (γ4) or Hb H (β4), visible as a precipitate in the red blood cells (Heinz's bodies) and identified during Hb electrophoresis;

- the α^0–thalassaemia with total absence of α-chain, lethal *in utero*.

The α–thalassaemias are essentially due to deletions affecting the α–globin locus and removing fragments of different sizes. One particular type of deletion is linked with the presence of repeated sequences surrounding the α–genes that promote unequal reciprocal recombinations, responsible for the loss of one α–gene on one of the chromatids and triplication of the gene on the other. This type of anomaly is represented by the two α–thalassaemia deletions '$-\alpha^{3.7}$' and '$-\alpha^{4.2}$' found in Africa and in south-east Asia respectively, and responsible for an α–thalassaemia phenotype ($-\alpha/\alpha\alpha$) in the heterozygous state. The triplications are silent unless they are associated with a β–thalassaemia trait that they make more severe.

The most frequently observed deletion takes a very large fragment, removing both α1– and α2–genes; it is called (α^0–thal $-^{sea}$) due to its frequency in south-east Asia, where it leads to α^0–thalassaemia ($-^{sea}/-^{sea}$) in the homozygous state, or to haemoglobin H disease in combination with the deletion ($-\alpha^{4.2}$), also very frequent in this part of the world (genotype : $-^{sea}/-\alpha^{4.2}$).

A certain number of point mutations in the α–chain are responsible for the α–thalassaemia phenotype, for example the mutation α Constant Spring α142 TAA$\rightarrow$CAA (stop$\rightarrow$Gln) that destabilizes the messenger RNA during its translation due to an increase in the length of the protein chain.

The methods for the molecular diagnosis of qualitative or quantitative mutations affecting the α–gene family are similar to those used for the β–gene family, with a predominance of the PCR technique used to identify deletions.

Dominant haemoglobinopathies

The large majority of the haemoglobin variants (missense mutations) are only responsible for severe pathologies when homozygous, but some are pathogenic in the heterozygous state.

The association, on the same allele, of the mutation β27 val$\rightarrow$Ileu with the classic sickle cell anaemia mutation β6 glu$\rightarrow$val (double-mutated allele) leads to the formation of haemoglobin HbS-Antille that can polymerize like the haemoglobin HbS in the heterozygous state despite the presence of a βA allele.

Some other mutants of the α– or β-chains are particularly unstable and are able to destabilize the tetramer and to produce severe chronic anaemias in the heterozygous state. These are often mutants affecting amino acids involved in making contact with the haem group.

2.3   Example diagnoses for X-linked diseases

2.3.1   Fragile X syndrome

Epidemiological, clinical and genetic aspects

Fragile X syndrome is the most frequent cause of inherited mental retardation, and the second most frequent cause overall of mental retardation after Down's syndrome. Its frequency in boys is 1 out of 4000 and in girls, 1 out of 7000. Among all of the genetic diseases called 'dynamic mutation diseases' or 'triplet repeat diseases' identified so far (see Table 2.3 and Section 2.4), fragile X syndrome was the first one to be identified, in 1991.

The disease is characterized by mental retardation, dysmorphic facial features and macro-orchidism in pubescent boys. When adult, the patients can perform manual work provided that they are taken charge of by relatives or by a suitable care structure. Fragile X syndrome has a variable expressivity, the degree of mental retardation varies from one individual to another, and the disease is often more discrete in girls.

Fragile X syndrome is characterized by a dominant transmission with an incomplete penetrance (the penetrance is the frequency with which the disease appears in individuals carrying the mutation). Only 80 per cent of men carrying a mutated chromosome are affected, which means that 20 per cent of male carriers are normal, and 55 per cent of female carriers are affected. In addition, the penetrance varies within one generation and it increases from one generation to the next (the phenomenon of genetic anticipation found in other 'triplet repeat diseases'). Therefore, the penetrance

Table 2.3   Dynamic mutation diseases

Disease	Triplet type	Number of triplets	
		Normal	Pathologic
Fragile X syndrome	$(CGG)n$	$n = 6\text{-}54$	$n \geq 200$
Kennedy's syndrome	$(CAG)n$	$n = 14\text{-}32$	$n \geq 40$
Steinert's myotonic dystrophy	$(CTG)n$	$n = 5\text{-}37$	$n \geq 50$
Huntington's disease	$(CAG)n$	$n = 6\text{-}35$	$n \geq 37$
Spinocerebellar ataxia type 1	$(CAG)n$	$n = 6\text{-}39$	$n \geq 40$
FRAXE mental retardation	$(CCG)n$	$n = 4\text{-}39$	$n \geq 200$
Jacobsen syndrome (FRA11B)	$(CGG)n$	$n = 11$	$n \geq 100$
DRPLA*	$(CAG)n$	$n = 3\text{-}36$	$n \geq 49$
Machado–Joseph disease (SCA3)	$(CAG)n$	$n = 13\text{-}44$	$n \geq 60$
Friedreich's ataxia	$(GAA)n$	$n = 6\text{-}29$	$n \geq 200$
Spinocerebellar ataxia type 7(SCA7)	$(CAG)n$	$n = 7\text{-}35$	$n \geq 37$
Spinocerebellar ataxia type 2(SCA2)	$(CAG)n$	$n = 13\text{-}32$	$n \geq 33$
Spinocerebellar ataxia type 6(SCA6)	$(CAG)n$	$n = 4\text{-}18$	$n \geq 21$

* Dentatorubral and pallidoluysian atrophy.

is strong in the brothers of an affected boy, but weak in the brothers of a boy who is a healthy carrier. It has also been observed that the daughters of male healthy carriers or affected men never have mental retardation, meaning that the mutation has to be passed through a female meiosis to be expressed in the child. This phenomenon of preferential transmission by one sex (called parental imprinting) is also found in many others diseases due to dynamic mutations, but not always in such a strict way. In the case of fragile X, it is probably due to contractions of the $(CGG)n$ sequence in the male germ line. These transmission particularities of fragile X posed some problems for prenatal diagnosis and genetic counselling until molecular biology provided the explanation for this surprising transmission, as well as tools allowing a reliable genetic diagnosis for the disease.

Molecular biology and physiopathology

Mutations responsible for fragile X syndrome   The 'fragile X mutation' is an amplification of a triplet Cytosine–Guanine–Guanine (CGG) repeat sequence, located in the 5′UTR of the first exon of the fragile X mental retardation 1 (FMR1) gene (Figure 2.11). The significant lengthening of this sequence due to the increase in the number of repeats is linked with a methylation of FMR1, which is then is no longer expressed. A normal chromosome has this repeat sequence $(CGG)n$, but with a low number of repeats, from six to 54 depending on the individual (Figure 2.11). In affected individuals, however, the number of repeats is over 200 and can reach 1000. Between 54 and 200 repeats, the carriers are said to be premutated: the FMR1 gene is normally expressed but the $(CGG)n$ sequence is very unstable leading to the complete mutation in subsequent generations. Besides the variability in the number of repetitions observed from one individual to another, both in normal individuals

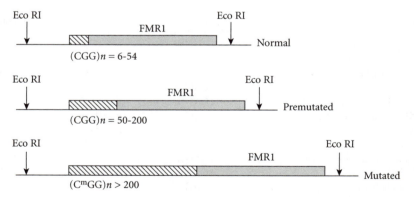

Schematic representation of the FMR1 gene and of the $(CGG)n$ triplet whose expansion is responsible for the fragile X syndrome. In the case of a complete mutation ($n>200$), the cytosines belonging to Cytosine–Guanine doublets (CmG) become methylated which represses gene expression.

Figure 2.11   FMR1 gene and CGG triplet

Table 2.4 Risk of transmission of a complete mutation by a pre-mutated mother depending on the number of CGG repeats – the transmitted chromosome is supposed to be the mutated one (from Fu *et al.*, 1991)

Size of the premutation (number of repeats)	Probability of transmission of a complete mutation
<60	<1 %
61–70	17 %
71–80	71 %
81–90	82 %
>91	99 %

and in patients, there is a variability within one individual, premutated or (mainly) mutated, with many different values for the number of repeats in different cells, indicating that this sequence is highly unstable during zygotic development. This somatic instability probably occurs very early during development. Because it abolishes FMR1 gene expression, it is the methylation that is responsible for the fragile X phenotype. Therefore, there is no correlation between the severity of mental retardation and the average size of the complete mutation at the level of the lymphocytes, which can actually vary slightly during life. It has been suggested that the size of the normal allele or of the premutated allele could have a role in intellectual functions, and notably learning ability, but this has not been confirmed.

The very special genetics of fragile X syndrome has a precise molecular basis that has begun to be well understood, even though some shadowy areas remain. The study of microsatellite markers in the fragile X region allows the identification of normal chromosomes carrying lengthened $(CGG)n$ sequences which point to the existence of a small number of 'founder' chromosomes that have given rise to the fragile X chromosomes (Oudet *et al.*, 1993).

The size of the CGG sequence is directly responsible for the status – normal, premutated or mutated (Fu *et al.*, 1991). The study of many pedigrees has also shown that it is the size of the premutation in the mother that determines the transfer of the complete mutation to her children, which leads to the phenomenon of anticipation (found in all genetic diseases with dynamic mutations). Indeed, the longer the premutation in the mother, the higher is the risk for her child carrying the complete mutation. If the size is below 60 repeats, she has no risk of having an affected child, but if it is above 90 repeats, the risk of having an affected child is 1 if the chromosome carrying the amplification is transmitted (Table 2.4).

The FMR1 gene   The FMR1 gene is expressed at different levels in many tissues, notably in the brain, testes and lymphocytes. It has been highly conserved through evolution, and codes for a protein involved in messenger RNA export from the nucleus to the cytoplasm (Eberhart *et al.*, 1996). The fact that the FMR1 protein binds different types of messenger RNA explains the pleiotropic effect of the fragile X mutation. The lack of gene expression due to methylation, associated with the lengthening of the (CGG)n sequence, constitutes most of the cases found in patients affected with fragile X syndrome, but a few cases of patients carrying point mutations leading to a loss of function in the FMR1 gene or a deletion of this gene have been described.

Molecular mechanisms leading to fragile X mutations   There are two distinct phenomena: the first one is an amplification, sometimes impressive, of the (CGG)n sequence which is probably responsible for a local difficulty in DNA replication, leading to a delay in replication and a more or less clean break, visible using cytogenetics; the second is a methylation of this region. The methylation might be a response to the amplification of the Cytosine–Guanine doublets present in each CGG triplet.

The molecular mechanism that could explain the (CGG)n sequence amplification is replication slippage. It is a phenomenon that occurs during DNA replication, essentially with di-nucleotide or tri-nucleotide repeated sequences, and that favours the formation of secondary structures on the single-stranded DNA due to Cytosine–Guanine pairing (Figure 2.12).

The size of the (CGG)n sequence and the existence or not of an interruption of the repeated motif by an AGG triplet are determining factors in the formation of those structures and the instability of the (CGG)n sequence (Hirst, Grawal and Davies, 1994; Eichler *et al.*, 1994). Some mutations in other genes (notably the genes involved in mismatch repair) could also play a role in (CGG)n sequence instability.

The genetic diagnosis

Cytogenetic diagnosis   The diagnosis of fragile X syndrome depended for a long time on the detection of the fragile site at Xq27.3 using cytogenetics, which is the origin of the name of this disease (Figure 2.13). The karyotype must be done in special lymphocyte culture conditions (media low in folic acid and with methotrexate). This cytogenetic diagnosis is made difficult by the fact that the fragile site can only be seen in 10 to 50 per cent of mitoses in affected boys and only 5 to 15 per cent of mitoses in affected girls. Also, about 50 per cent of female carriers do not show the fragile site. There is no correlation between the level of expression of the fragile site (that is to say the percentage of mitoses where the extremity of the X chromosome is broken) and the severity of the disease, in either boys or girls.

Molecular diagnosis   Even though a few point mutations and deletions of the FMR1 gene have been described in the literature, most of the fragile X cases are due to (CGG)n sequence amplification.

Molecular probes for the FMR1 gene give a very satisfactory system for genetic diagnosis of the disease because they make it possible to distinguish all the possible statuses (normal, premutated and mutated) in both sexes. The method relies on the Southern technique and the use of two restriction enzymes, one of them being sensitive to the methylation state of the gene and allowing the detection of the

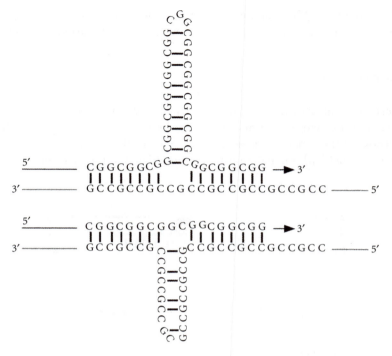

(a) If the hairpin forms on the newly-synthesized strand and if the S-phase of the next cell cycle starts before any reparation is carried out, the elongated newly synthesized strand will be used as a template to form a DNA duplex carrying the amplified sequence. In the case where there is reparation before the S-phase starts, the original sequence will be restored.

(b) If the hairpin forms on the template strand and if the S-phase of the next cell cycle starts before any reparation is carried out, the shortened newly synthesized strand will be used as a template to form a DNA duplex carrying the shortened sequence. In the case where there is reparation before the S-phase starts, the original sequence will not be restored and will stay contracted.

Figure 2.12   Replication slippage:this occurs when 'hairpin' secondary structures, stabilized by hydrogen bonds, form during DNA synthesis between cytosine and guanine nucleotides of one strand

methylation associated with the complete mutation (Figure 2.13). This method is reliable and can give the result of the diagnosis in 8 to 15 days. Most of the laboratories have also developed a method using PCR to measure precisely on an electrophoresis gel the size of the small premutations, an important parameter for the prediction of $(CGG)n$ sequence stability (Figure 2.14) and for the determination of the carrier or non-carrier status of the relatives of the patient. This method has the advantage of being cheaper and faster than the Southern method and is also used to search for the fragile X mutation in patients suffering from mental retardation. However, the large number of CGG triplets makes amplification of the mutated alleles difficult. The PCR is therefore a useful complementary tool but it cannot completely replace the Southern technique, notably in girls in whom it is not possible to distinguish a normal allele in the homozygous state from a normal allele associated with a mutated allele that could not be amplified by PCR. The PCR makes it possible to recognize quickly all unaffected boys and 65 per cent of unaffected girls (the average rate of heterozygosity for the (CGG)n sequence in the general population is 65 per cent).

Genetic counselling

In a certain number of families, the diagnosis of fragile X syndrome has already been made using cytogenetics and molecular biology. Therefore, most of the time, the purpose of the genetic counselling is to confirm the existence of a mutation in a patient affected with mental retardation and related to the proband, or to search

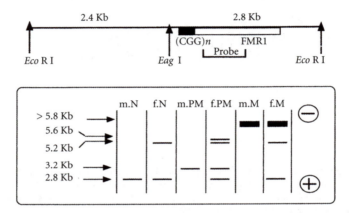

(a) Restriction profiles after double digest (enzymes *Eco* RI and *Eag* I) of the genomic DNA and hybridization with a molecular probe recognizing the FMR1 gene. The Eag I enzyme is sensitive to methylation and does not hydrolyse methylated DNA, notably the inactive X in females
- m.N. Normal male
- f.N. Normal female: the 5.2 kb fragment corresponds to the sequence on the methylated inactive X chromosome
- m.PM. Premutated male with an amplification of about 400 bp = 130 CGG triplets
- f.PM Premutated female: the four restriction fragments observed correspond to the four possible types of chromosome: normal active X (2.8 kb), normal inactive X (5.2 kb), premutated active X (3.2 kb) and premutated inactive X (5.6 kb).
- m.M. Mutated male: male carrying a complete mutation. The methylation associated with the mutation leads to the absence of restriction by the enzyme Eag I. The smear reflects the variability in the size increase of the CGG sequence in the analysed cells (leucocytes for example)
- f.M. Mutated female: in addition to the complete mutation, two fragments of 2.8 and 5.2 kb corresponding to the two possible statuses (active and inactive) of the normal chromosome X are visible.
Note that the complete mutation and the premutation are co-existing in 10 to 15 per cent of patients
(b) Autoradiography of a Southern blot
The genomic DNA was digested with the enzymes *Eco* RI and *Eag* I, separated by gel electrophoresis, transferred onto a nylon membrane and hybridized with the molecular probe pFxa 1NHE labelled with digoxygenin.
M-DNA marker (lambda bacteriophage DNA hydrolysed by the enzyme HindIII: see Chapter I)
1: Female carrier for a premutation; 2: male carrying a complete mutation; 3 and 4: normal males; 5 and 6: normal females.

Figure 2.13   Fragile X syndrome diagnosis using the Southern technique

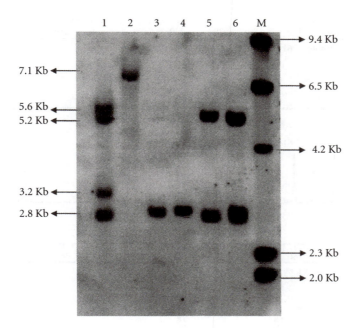

Figure 2.13 *(Continued)*

if a woman of childbearing age is a carrier. A second common case consists of searching for the fragile X mutation in a child suffering from mental retardation or behavioural problems. Finally, in a few cases, the fragile X mutation is searched for because cytogenetics reveals a break on the long arm of the X chromosome when performing a standard karyotype for a patient with mental retardation. Fragile X syndrome has also been identified in a few families thanks to the existence of a girl affected with a moderate mental retardation.

When the anomaly is already known in the family, a prenatal diagnosis can be offered to female carriers, normally from the second month of pregnancy and using a chorionic villus biopsy. A karyotype revealing the sex is performed, and the fragile X mutation is searched for using the Southern technique and PCR. The same tests can be performed using amniotic cell cultures from amniosynthesis. Pre-implantation diagnosis is currently being developed.

2.3.2 Genetic diagnosis of haemophilia A and B

Epidemiologic and clinical aspects

Haemophilias are X-linked, recessive, hereditary bleeding disorders, and are caused by a deficiency in clotting factor VIII (haemophilia A) or in clotting factor IX (haemophilia B). Haemophilia A, that concerns one male birth out of 5000, is more

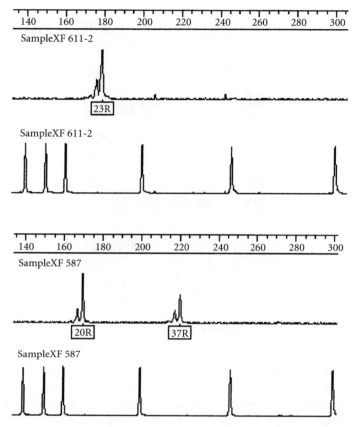

The PCR products are labelled with fluorescent bases detectable by the light emitted when they are excited by a laser beam on passing through the read window at the bottom of the gel in an automated sequencer. The co-migration of a DNA standard allows the determination of the size of the PCR products, and therefore the identification of the alleles. Top: normal male carrying an allele with 23 CGG repeats. Bottom: normal female with two alleles carrying 20 and 37 CGG repeats.

Figure 2.14   PCR analysis of the (CGG)n sequence responsible for fragile X

frequent than haemophilia B (about one male birth out of 30 000). The clinical severity depends on the importance of the deficiency, giving several forms: severe haemophilia, with a residual level of activity for the factor VIII or IX less than or equal to 1 per cent, moderate haemophilia, when the residual level is between 2 and 5 per cent and the minor forms with a residual level between 6 and 30 per cent. Prevention and treatment of bleeding accidents can nowadays be kept more or less under control, and life-threatening events are essentially linked with the complications of the substitutive treatment by transfusion: virus contaminations, inhibitory antibodies against one of the transfused factors that cancel the effect of the treatment. Therefore, there are many families asking for genetic counselling and a prenatal diagnosis can be requested.

The main aim of the genetic diagnosis of haemophilias is to identify women who are carriers of haemophilia, meaning the women are heterozygous, who are not affected but are able to transmit the affected allele, and therefore the disease, to their sons. In theory, when it is a sporadic haemophilia, a third of the cases are due to new mutations (see Haldane's rule, Chapter 7). In this case, the purpose of the genetic study is to define with certainty the status of the women requesting, in order to perform eventually an early prenatal diagnosis for the carriers.

Biochemical and genetic aspects

Clotting factors VIII and IX   The gene coding for clotting factor VIII (F8) is one of the 'big' genes, covering 186 kilobases of the long arm of the X chromosome (position Xq28). It has 26 exons coding for a protein having 2351 amino acids. The release of the signal peptide allows the secretion of a still inactive protein, made up of three types of domains A,B and C, more or less repeated. In the bloodstream, it forms a complex with the von Willebrand factor (FvW). Activation by thrombin leads to the complex dissociating, and starts a cascade of protein cleavages, the produced fragments associating through calcium ions to form the activated factor VIII.

The gene for clotting factor IX (F9) is of medium size (33 kilobases), located at position Xq27 and contains eight exons. The mRNA is translated into a pro-peptide containing six different functional domains. The cleavage of the signal peptide, and then of the following 18 amino acids, generates a mature protein of 415 amino acids. Activation is the result of the excision of a central peptide, generating a light chain and a heavy chain that stay linked through a disulphide bond. The activated factors VIII and IX allow the activation of factor X and, therefore, have a central role in the intrinsic coagulation pathway.

Molecular pathology of haemophilia A   A recurrent mutational mechanism accounts for about 45 per cent of the severe cases of haemophilia A. It is an inversion of one segment of the X chromosome containing exons 1 to 22 of the F8 gene. This inversion is the result of an intrachromosomal recombination between these two homologous sequences, one located in intron 22 and the other one downstream of the gene, towards the telomere of the long arm of chromosome X (Figure 2.15), and seems to happen essentially during male meiosis. The existence, upstream of the gene, of several copies homologous to the one in intron 22, is responsible for several types of inversion. In all cases, the result is an interruption of the gene that cannot be transcribed entirely. More recently, another recurrent inversion, much rarer, has been characterized at the level of intron 1.

Many point mutations have been described, scattered all over the coding sequence. They consist of 'private' mutations, meaning that they are different in each family because there is no recurrent mutational mechanism. These mutations account for about 50 per cent of the severe cases of haemophilia A and almost all of the moderate

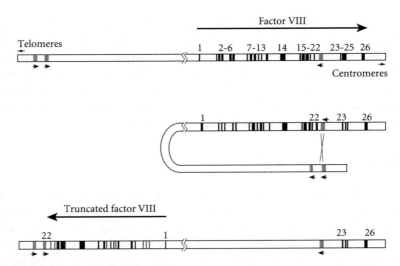

The existence of homologous sequences between intron 22 of the gene and a telomeric region (grey boxes) favours a pathologic intrachromosomal pairing during meiosis. The recombination between intron 22 and one of the telomeric sequences leads to the inversion of a part of the X chromosome, not detectable through cytogenetics, which interrupts the F8 gene. This gene can be expressed as a truncated protein corresponding to exons 1 to 22 (from Lakich *et al.*, 1993)

Figure 2.15   Inversion mechanisms in the F8 gene

or minor cases. They can be no-sense mutations (stop codon) responsible for severe forms, missense mutations with a variable severity, and mutations causing splicing defects, also associated with severe, moderate or minor forms. Deletions/insertions of a few base pairs can also cause haemophilias of variable severity depending on whether or not they cause a frameshift.

The remaining 5 per cent of severe haemophilia cases are due to rearrangements, most of the time large deletions covering different regions of the gene but whose breakpoints are not determined in most cases. Other mutational mechanisms have been exceptionally described, like duplications of certain exons, insertions of repeat sequences or X-autosomal translocations. The phenotypic variability that can be observed for a particular mutation can be explained by the influence of additional mutations in the F8 gene or by the presence of other coagulation anomalies. The appearance of inhibitors (antibodies against the transfused factor VIII) occurs most of the time in severe forms, but is described in a few moderate forms associated with no-sense mutations.

Molecular pathology of haemophilia B   Most of the mutations (95 per cent) are no-sense or missense point mutations, insertions/deletions of a few base pairs. Several hundred different mutations are reported and are scattered all over the gene. The remaining 5 per cent are due to larger deletions.

Diagnosis of haemophilias

Biological diagnosis The biological diagnosis relies on the dosage of the clotting activities of the two factors (FVIIIc or FIXc) in haemophiliacs. The combination with an antigen dosage makes it possible to distinguish between haemophilia where the deficient factor is absent in the bloodstream (levels of clotting factors and antigens decreased) from haemophilia where the factor is present but not functional (the level of clotting factor diminished but antigens normal or slightly diminished). The dosage of the clotting factors in women for whom the carrier status is being determined is not always informative: because of the lyonization process (random X chromosome inactivation in women) and of the high variability of the normal values of the clotting factors, a third of the carrier women have a normal haemostasis assessment.

The detection of haemophilia B carriers relies on the dosage of FIXc: a dosage reading of FIXc < 0.6 several times in a female relative of a patient with haemophilia B is sufficient to diagnose her as a carrier. Concerning the detection of haemophilia A, the dosage of factor VIIIc is not sufficient because its level also depends on the FvW (its transporter) level in the bloodstream, Therefore, it is necessary to measure the FvW antigen and to calculate the fraction FVIIIc/FvWAg. These readings should not be performed during pregnancy as the synthesis of these factors in the liver increases at this time. A reading showing the fraction FVIIIc/FvWAg < 0.7 several times in a female relative is sufficient to diagnose her as a carrier.

Molecular diagnosis In general, two types of approach allow the detection of a mutated allele: a direct approach, based on the identification of the mutations responsible for the disease, and an indirect approach relying on the linkage between the disease locus and polymorphic markers located near or inside the gene. The advantage of the direct analysis is considerable in haemophilia where the status of the women has to be determined. The identification of a pathogenic mutation in a haemophiliac makes it possible to offer a simple screening of all potential carrier women in the family, even for those who are only distantly related. In contrast, the indirect approach often requires a study of a large number of family members and cannot give the status of the mother of a sporadic haemophiliac (the mothers of affected individuals are not necessarily carriers because one case out of three can result from a new mutation, see Chapter 7).

The direct approach for moderate haemophilia A relies on the search for point mutations in all coding sequences of the gene. This work is time consuming because of the size of the gene and the heterogeneity of the mutations, and can only be done by a few laboratories. Therefore, an indirect approach is more often used (see below).

The direct approach is more often proposed in severe forms for the search of an inversion. This is usually performed on the genomic DNA by Southern blot, which can detect several inversion types (Figure 2.16). About 85 per cent of the inversions are due to a recombination between a sequence in intron 22 of the *F*8 gene and an extragenic homologous copy located distally (type 1 inversion), 12 per cent are due

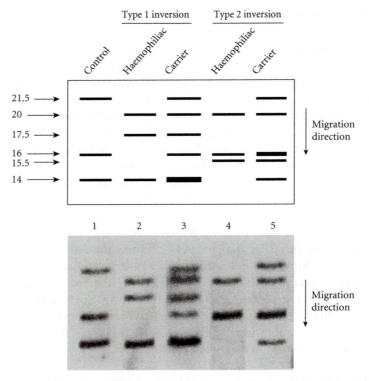

The two most frequent types of inversion are represented in this figure and the corresponding photo. Hybridization is realized with a probe (probe a) corresponding to the part of intron 22 of the gene F8 whose sequence is also present outside the gene and is responsible for inversions (see Figure 2.15).
The 21.5 kb band corresponds to the intragenic copy (intron 22), the 16 kb band corresponds to the distal extragenic copy and the 14kb one, to the proximal extragenic copy. The 20, 17.5 and 15.5 kb bands correspond to copies rearranged by the inversion.

Figure 2.16   Southern blot to search for an inversion of the F8 gene

to a recombination with a proximal homologous copy (type 2 inversion) and the remaining 3 per cent result from a recombination with an extra copy. More recently, the setting up of a long fragment amplification technique (long PCR) facilitates the search for inversion, but cannot determine the type of inversion involved.

The indirect approach relies on the study of the co-transmission of the pathogenic mutation with DNA markers. Several types of markers are used.

- RFLPs (di-allelic, see Chapter 1), notably a BclI site in intron 18, and an XbaI site in intron 22. In the latter case, the presence of repeat sequences in intron 22 and upstream of the gene (see previous paragraph) makes the interpretation of the results more complicated.

- Microsatellites (see Chapter 1), notably at the level of introns 13 and 22, the most used markers because they are multi-allelic (Figure 2.17).

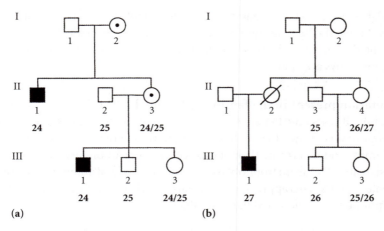

(a) According to the data given by the pedigree, I2 and II3 are carriers. The III3 female inherited from her mother II3 the allele 24 that is linked with haemophilia in this family. Therefore she can be considered as a carrier for haemophilia A, assuming there is no recombination (the risk of which is generally considered to be negligible with an intragenic marker). It has to be noted that in the case represented here, a false paternity can lead to a diagnostic error in III3.

(b) This example represents a sporadic case of haemophilia. The female III3 does not have a common allele with the haemophiliac III1 (exclusion diagnosis). However, it is not possible to conclude on the status of II4 because she shares an allele with her haemophiliac nephew.

Figure 2.17   Indirect diagnosis example using intragenic microsatellites analysis

When the intragenic markers do not give any information, it is possible to use extragenic markers but it is indispensable to take into account the risk of having a recombination between the marker and the $F8$gene that can cause an error in the diagnosis. The development of genetic and physical genome maps during the last 10 years makes it possible to study several markers located on both side of the factor VIII gene, in order to detect recombinations and to give a result with a satisfactory security.

In the case of haemophilia B, it is simpler because the smaller size of the gene makes the direct study more accessible as a routine technique, despite the large amount of molecular heterogeneity. It consists of the use of 'scanning' techniques to test the whole gene for the presence of mutations responsible for the disease in each family. In the case of a negative result, large deletions can be searched for. Finally, several intragenic markers allow realization of an indirect diagnosis, in the very few cases where no mutations have been detected using the direct approach.

Genetic counselling

Its purpose is to answer the following two questions.

- Is a female relative a carrier of the anomaly?
- In the case of a severe haemophilia, can she be offered a prenatal diagnosis?

Molecular genetics makes it possible to answer both questions in a large number of cases. Prenatal diagnosis always starts with the determination of the sex of the fetus using the karyotype because, except in a few cases (see paragraph 'special cases' below), only boys are affected.

The mutation present in the family is identified   This is the case in almost all haemophilia B cases and 45 per cent of severe haemophilia A cases due to inversion. It is therefore very simple to search for the mutation in the potential female carriers in the family, and determine their status without any ambiguity. An early prenatal diagnosis using DNA from trophoblast cells after 12 weeks of amenorrhea is offered to the female carriers. This approach can be particularly useful to solve certain difficult cases (sporadic cases, deceased haemophiliac).

The causal mutation cannot be identified   The test using DNA markers is the only one possible; consequently, each case is special. The determination of female carrier status depends on their relationship with the haemophiliac and on the informativity of the markers.

The situation is particularly complicated when it concerns sporadic haemophilia. A woman carrying alleles other than the alleles found in the related haemophiliac for intragenic markers could be reassured ('exclusion' diagnosis, Figure 2.17). In contrast, if she carries alleles in common with the related haemophiliac, no conclusion can be made because it is not possible to know when the mutation appeared in the family – the mutation could indeed have been silently transmitted by female carriers during several generations. In this case, a direct analysis through the search for point mutations in the gene coding for factor VIII is highly recommended, but it is important to keep in mind that such an approach can take longer because it is generally part of a clinical research study.

For a prenatal diagnosis using the indirect approach to be offered to a woman shown to be a carrier (according to her family history, haemostasis tests or a molecular study), she must be heterozygous for at least one of the markers.

Molecular study cannot give an answer   In this case, the risk of a woman being a carrier should be precisely calculated, taking into account her position in the pedigree, the result of her haemostasis tests and whether or not she previously has had healthy sons (theory of conditional probability).

Ultimately, when the prenatal diagnosis cannot be performed using molecular genetics, it can be done by measuring the level of clotting factors in blood from the umbilical cord, after a sonogram has been performed to determine the sex. The collection and the dosage are delicate. The contamination of the cord blood with maternal blood or by amniotic fluid can be at the origin of errors in the diagnosis and requires checking the purity of the fetal collection.

Special cases Girls can exceptionally be affected with haemophilia. The causes are:

- skewed inactivation of the X chromosome carrying the normal allele (the expressed gene is the one carrying the mutation);

- Turner syndrome 45, X;

- translocation between an X-chromosome and an autosome;

- X chromosome isodisomy (nondisjunction of the X chromosome during the second maternal meiotic division, responsible for the presence of two identical X chromosomes in the zygote).

There is also a particular from of von Willebrand disease, leading to a binding anomaly between factor VIII and the von Willebrand factor, which simulates a haemophilia clinical picture. The transmission is autosomal recessive and affects girls as well as boys.

Conclusion

Haemophilia A and B are two relatively frequent X linked hereditary diseases, that can still justify, mainly for the severe forms, the request for a prenatal diagnosis. Female carriers are still very difficult to diagnose. Knowing the molecular defects at the origin of the disease should make this less difficult and allow the determination of the carrier status of these women to offer an early prenatal diagnosis. As for all genetic diseases, the direct study of the causative mutation in each family should be preferred when possible. In contrast, the indirect approach through the study of marker segregation makes it possible to solve most of the cases, but the study of sporadic haemophilia A can sometimes still remain difficult.

2.3.3 *Molecular diagnosis of Duchenne and Becker muscular dystrophies*

Mode of transmission and clinical feature

Duchenne and Becker muscular dystrophies are inherited as X-linked recessive traits. As for haemophilia, only men are affected and women are carriers of the disease, showing signs only exceptionally. However, there are women really affected by Duchenne muscular dystrophy due to homozygosity or hemizygosity,

Duchenne muscular dystrophy is the most severe of the myopathies. The prevalence at birth is about 1 out of 3 500 male newborns. The disease was identified by Dr Guillaume Duchenne in 1860. The first clinical signs of the disease are subtle and can often stay undetected for several years. A small delay in learning to walk and frequent falls can sometimes be noticed. The clinical picture is more precise around 3 to 4 years old: the child can have difficulty climbing the stairs, or getting up from a prone position. The way of walking is duck-like. Running is almost impossible and falls are frequent. When fallen, the child can only get up by climbing along his legs (Gower sign). At this stage, the clinical examination shows a hypertrophia of the calves in 80 per cent of the cases. The level of creatine kinases (CK, a muscle enzyme) in the serum is very high. The evolution of the disease is then unavoidable and corresponds to a muscular atrophy affecting every muscle progressively. Walking becomes harder and harder and at the age of 10, it is only possible using orthotic devices. At about 12 years old the child cannot walk anymore. When the muscles of the spine and of the upper limbs are affected it leads to spinal deformation and thorax deformation causing respiratory insufficiency. In parallel, retraction of the muscles and tendons of the lower limbs occurs with flexion of the hips and knees, and equinism of the feet. At about 15–16 years old, the patient becomes completely dependant. Very often a myocardiopathy occurs. Death happens between 16 and 25 years old with several clinical pictures: respiratory failure, cardiac failure or paralytic ileus. Even if the effect on muscles is the best known among the defects, a third of those affected also suffer from a cognitive deficiency (average IQ of 80). The mental retardation is not progressive and is not correlated to the muscular defect. It leads to a decrease of the verbal intellectual capabilities, reading defects and problems with other memory functions.

Becker muscular dystrophy was identified in 1955. It consists of a muscular dystrophy evolving the same way as the Duchenne muscular dystrophy, but much more slowly. The first clinical signs appear between 5 and 25 years old. There is also pseudo hypertrophy of the calf muscles, but the muscle weakness is less severe in general. The life expectancy of the patient is longer, but after 40, the heart should be tested regularly. This myopathy is rarer, with an estimated prevalence at birth of 1 out of 18 500 male newborns.

It must be noted that two thirds of female carriers for muscular dystrophy have an elevated level of CK, and 17 per cent present muscular signs, even though they are normally considered as being not affected. Several studies also showed that 40 per cent of female carriers over 16 have myocardial defects such as hypertrophy (30 per cent) or dilatation (10 per cent). These observations justify a systematic cardiac examination but it is still carried out too rarely.

Electromyogram and muscle biopsy

In the Duchenne disease, the electromyogram reveals a pattern consistent with myopathy: no muscular activity at rest, shortening of the duration of the motor unit potentials and polyphasic potentials. There is also a contrast between the weakness of the voluntary movements and the richness of the collected traces.

This test – always pathological in affected individuals – also shows anomalies in some female carriers. The muscle biopsy is also a usual element for the diagnosis. It shows a defect in the diameter of the fibres (10 to 200 μm instead of 50 μm) and an anarchic distribution of atrophied fibres. There are also hypertrophied fibres. The connective tissue is increased and the endomysial fibrosis is intense. Within the muscular fibre, some nuclei appear to be hyperchromic with an increase in their size and number. They lose their peripheral position to migrate to the centre of the fibres. There is fibrillar degeneration with necrosis, phagocytosis and regeneration, together leading to a cellular impoverishment that can be extreme. All of these alterations appear progressively, giving an aspect that evolves with the disease stage. Finally, different levels of histological defect can be observed in the muscles of some female carriers, although there is no general pattern for this occurrence.

The gene and function of dystrophin

The Duchenne and Becker muscular dystrophies are 'allelic diseases' because they are associated with defects in the same gene: the dystrophin gene, located on the short arm of chromosome X (Xp21). It is the largest gene found so far in humans with a size of 2.4 Mb. It contains 79 exons and encodes a major transcript of 14 kb. The main product of the gene, dystrophin, has a molecular weight of 427 kDa. Dystrophin is a sarcoplasmic protein localized just under the sarcolemma (Figure 2.18). Dystrophin has the property of self-association like proteins from the actinin and spectrin families. Dystrophin molecules associate to form variable size filaments on which individual molecules can slide. The dystrophins could be the principal elements of an elastic net, or a kind of web inside the muscle cell. They could slide against each other allowing the deformation of the entire net during contraction and relaxation. The dystrophin interacts with two types of protein (Figure 2.18): first, with proteins of the sarcoplasma (actin, dystrobrevines, syntrophines) and second, with a complex of glycoproteins that are part of the membrane. These proteins associated with the dystrophin (dystrophin associated glycoproteins or DAG) permit its attachment to the sarcolemma. The DAG themselves interact with the laminin, a component of the extracellular matrix. Finally, the dystrophin gene has several promoters with different tissue specificities (lymphoblast, brain, cerebellum, retina, kidney, peripheral nervous system) that give rise to transcripts and proteins with different sizes and functions, still poorly characterized.

Molecular defect of the gene and molecular analysis methods

Deletions Systematic analysis of the DNA from individuals affected with Duchenne or Becker muscular dystrophy showed that partial deletions are the major cause for the gene dysfunction. The frequency of the observed deletions increased

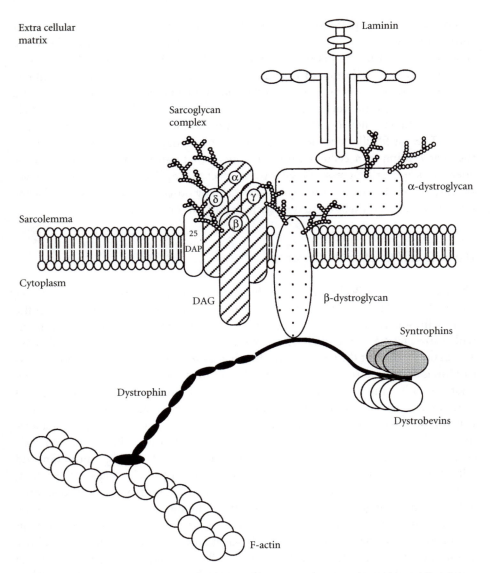

Figure 2.18   Model of interaction of dystrophin with cytoplasmic and sarcolemma proteins

during the process of cloning the gene (partial genomic probes, then total cDNA and entire gene cloning). The size of the deletions is very variable – it goes from 0.5 kb to 2000 kb, with rare deletions of the entire gene or even of the genes next to it. Studies undertaken by several groups showed that the deletions affect mostly two regions of the gene, called deletion hot spots. These two regions of predilection are located in the 3′ median region and in the 5′ region of the gene. The first hot spot

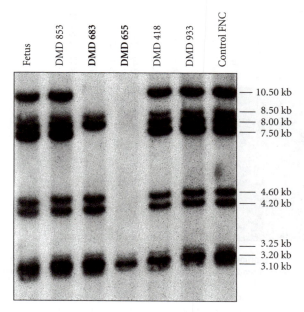

A Southern blot with HindIII fragments of genomic DNA was hybridized with the 9.7 probe that recognizes nine fragments of well-defined size corresponding to exons 9 to 11 of the dystrophin gene. A fetus, five cases and one control (FNC: female non-carrier) have been studied.

The comparison of the control's profile with those of the cases reveals the absence of bands corresponding to the presence of a deletion in DMD655 and DMD683. For DMD655, the only band left (3.2 kb) is specific to exon 1. This patient presents a deletion covering the exons 2 to 11. For individual DMD683, two bands are missing corresponding to a deletion of the exons 8 to 11.

Figure 2.19   Examples of deletions in the dystrophin gene as shown by Southern blot

concerns exons 43 to 52 with a frequent break point in intron 44. The frequency of the breakpoints in this intron is not perfectly understood yet. The large size of the intron (160 to 180 kb) is not sufficient reason to explain such a proportion (30 to 40 per cent depending on the authors). The search for repeat sequences has also not shown evidence that the effect is due to a high unequal recombination probability (Blonden *et al.*, 1991). The second deletion hot spot is in the 5′ region of the gene with breakpoints in intron 7 in 10 to 15 per cent of cases depending on the authors. The size of this intron is estimated to be 110 kb. Overall, the deletions detected in this region cover a domain larger than with the first region. Finally, it should be noted that deletions after the exon 60 are very rare.

The search for deletions requires use of the Southern method (Figure 2.19) and 'multiplex' PCR (Figure 2.20) techniques. In the latter case, several regions of the gene are amplified and examined simultaneously. The probes for the amplification products obtained within a single reaction are chosen to have different sizes to make it simple to distinguish them. In this way, 44 exons can be tested quickly using only four multiplex amplification reactions.

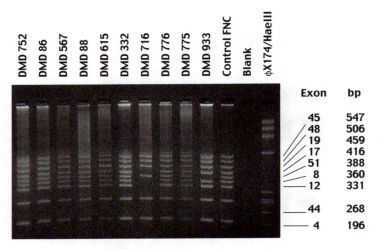

In this analysis, a set of 18 primers allows the simultaneous amplification of nine exons. The exon number and the size of the amplified product are indicated for each band on the right of the figure. The gel analysis reveals the presence of nine bands in the control (FNC: female non-carrier) and gives a quality control for the multiplex PCR due to the absence of any unspecific amplification in the 'blank' (tube without DNA). The analysis of each lane for each studied patient reveals the absence of a variable number of bands corresponding to different deletions. In fact, patient DMD933 is a carrier for a deletion covering exon 51, patient DMD571 is a carrier for a deletion that affects at least exons 12 to 17 but that does not expand upstream of exon 5 (presence of exon 4) or downstream of the exon 18 (exon 19 present), etc..

Figure 2.20   Examples of deletions in the dystrophin gene as shown by multiplex PCR techniques

Duplications   Some partial duplications have also been identified. The localization of these duplications is widely spread and variable. Less than a third of the duplications are found in the deletion hot spots, and there does not seem to be any duplication hot spot, even though they seem to be more concentrated in the 5′ region of the gene (exons 5–7). The duplication frequency is estimated at 6 to 7 per cent depending on the laboratory. This number is probably underestimated because they are difficult to identify. Indeed, the Southern and multiplex PCR methods used for the identification and analysis of deletions (see Figures 2.19 and 2.20) are useless to detect duplications because the hybridized or amplified sequences are present and because the functional alteration of the gene comes from the duplication of one of them, requiring the use of quantitative methods (Southern or real-time PCR, see Chapter 1). In some cases, the duplication can be observed thanks to the presence of an abnormal restriction fragment by the Southern method or pulsed field gel electrophoresis. Several mechanisms, such as unequal crossing over (sister chromatid exchange), intra-chromatid exchange, replication slippage (see fragile X syndrome in this chapter) can explain the appearance of a duplication.

Point mutations   About a third of the mutations responsible for the Duchenne and Becker muscular dystrophies are associated with point mutations or with very small alterations of the dystrophin gene, that cannot be detected by the Southern technique or by PCR. The international database (http://www.dmd.nl) has at the moment information on 1588 unique point-mutations. These mutations are essentially no-sense mutations, rarely recurrent and spread all over the gene, without any mutation hot spot. However, because almost all these mutations lead to truncated proteins, they can be search for using a test specially adapted to this mutational context: the protein truncation test (PTT, see Chapter 1). By combining an RT–PCR and an *in vitro* transcription translation system, it is possible to visualize the abnormal protein product. Several protocols and several pairs of primers have been reported. In general, 10 reactions are necessary to test the conformity of the complete coding sequence of the gene.

A special context concerning mental deficiency found in one third of patients suffering from muscular dystrophy should be mentioned. Four transcripts of the dystrophin in the brain are known: two full-size isoforms and two isoforms that are shorter distally Dp140 (promoter in intron 433) and Dp71 (exon 63–79). There is a preferential link between cognitive defects and mutations in the dystrophin gene. It consists of distal mutations (starting from exon 40), with more severe effects associated with mutations located between the exons 55 and 72.

Genotype/phenotype correlations and physiopathology

Genotype/phenotype correlations have been studied since the first deletions were identified. At that time, however, the phenotype severity was not explained by the localization or the size of the deletion. As soon as anti-dystrophin antibodies were available, immunohistological studies revealed different results between patients affected with the Duchenne type muscular dystrophy, and patients affected with the Becker type: for the patients affected with the Duchenne type, almost no labelling could be seen while for the Becker's patients, there was a signal but very weak. The same applied for the Western blot analysis: the protein was totally absent for the Duchenne's patients and a smaller immuno-reactive product was observed in the Becker's patients. In 1988, Monaco *et al.* proposed a molecular mechanism to explain this difference. An intragenic deletion can have two consequences: it can create a frameshift leading to a wrong transcript that cannot give a protein product, or the deletion can have no effect on the reading frame and therefore induces the synthesis of a smaller protein. Later, the precise borders of the exon/intron boundaries in the dystrophin gene confirmed the 'reading frame' theory and its different effects on the severity and the evolution of the Duchenne and Becker types (Roberts *et al.*, 1993). The identification of the point mutations also gave results in accordance with this theory. Indeed, in 98 per cent of the Duchenne's patients, the mutation creates an early stop codon, while splicing mutations are more frequent in the Becker's patients. However, in about 8 per cent of the cases, there is a discrepancy between the reading frame and the disease severity, with some Becker's patients suffering from a severe form and some Duchenne's patients not showing any frameshift. The most typical exception is a deletion that covers exons 3 to 7 and creates a frameshift leading to a Duchenne or to a Becker type muscular dystrophy. Many antibodies directed against different epitopes of the dystrophin are available now. These antibodies are routinely used for the diagnosis of the Duchenne and Becker forms. They can also be used in some circumstances to perform a prenatal diagnosis on a fetal muscle biopsy.

General instability of the dystrophin gene and intragenic recombinations

From the first stages of the cloning of the dystrophin gene, a relatively high recombination level was observed, suggesting that the gene could be quite large. However, even before its size was known, there was a discrepancy between the genetic size (expressed in terms of the recombination frequency) and the physical size of the cloned region. For a distance of about 2.4 Mb, the theoretical risk of recombination should be close to 2.5 per cent. In fact, a much higher recombination rate was reported by different groups, and is between 7 and 10 per cent for the entire gene. A detailed study of the gene shows that the recombination rate is not homogenous over the entire gene. Therefore, in the same way that there are two deletion hot spots, there are two recombination hot spots, the two being more or less identical. The first region is at the 5′ end of the gene, between introns 1 and 8, and the second one is centred on introns 44 and 45. In these regions, the recombination rates are 7 and 5 per cent respectively. Finally, some true double recombinants have been identified without any ambiguity making these types of event not just a hypothetical possibility for the dystrophin gene, and making family analysis using polymorphic markers more difficult.

Mosaicism

Duchenne muscular dystrophy is a lethal genetic disease with a high frequency. The geneticist Haldane showed in the 1930s that a third of patients affected with an X-linked disease were carriers of a *de novo* mutation (see Chapter 7). The high frequency of the Duchenne and Becker muscular dystrophies means that a high spontaneous mutation rate affects the dystrophin gene, which is linked with the two recombination/deletion hot spots identified and the fact that most of the mutations are precisely deletions or duplications. This phenomenon is confirmed by the observation of families in which the patients suffering from dystrophy carry an identical family mutation except for one of them who has a different mutation, which is a new mutation. In addition, abnormal recurrence of the disease has been identified, and this event could be explained by a phenomenon of germ cells mosaicism of a paternal or maternal origin. This phenomenon makes the evaluation of the status of at risk women more complicated in some circumstances, such as when the family mutation is unknown and the diagnosis has to be made by indirect family analysis.

Strategies for biological diagnosis (limits and reliability, efficiency)

Study of dystrophin and of its gene had three main purposes: a precise proband diagnosis, the evaluation of the status of at risk women and prenatal diagnosis.

Proband diagnosis   The study of dystrophin and the study of its gene will be seen in two situations, often not independent of each other: to confirm the Duchenne–Becker clinical diagnosis at the molecular level, and to identify precisely the molecular defect in the context of a family study, notably to prepare for a prenatal diagnosis.

Protein analysis. Dystrophin analysis of a muscular biopsy is of major importance to distinguish a dystrophinopathy from an autosomal recessive form of dystrophy and therefore orientate the molecular study. Initially limited to the analysis of dystrophin using two specific antibodies, the analyses are now performed with eight antibodies allowing the study of six proteins: dystrophin, sarcoglycans (α, β, γ, δ) and calpain. There is a multiplex system of Western blotting that allows the simultaneous study of the six proteins. This fast method also allows the detection of size differences, even if minimal, like the deletion of just exon 48. This technique is expensive, delicate to perform and requires a high level of expertise. Also, the results can be hard to interpret due to a large variation in the quantities of sarcoglycans found in the Duchenne's patients (from zero to an almost normal quantity) or in the quantity of dystrophin in the patients suffering from 'adhalinopathy' (DAG2 or α–sarcoglycan alteration, see Figure 2.18) and not from a Becker dystrophy.

Search for deletion and duplication of the gene. The study will be performed most of the time using lymphocyte genomic DNA extracted from a blood sample. The strategy used consists of searching first for a deletion in one of the mutation hot spots using gene amplification. Once the region of the deletion has been identified, its exact size (in terms of the exact number of exons deleted) will be determined either by amplification of the flanking exons or by the Southern technique. Next, several blots will be obtained and they will be hybridized successively with different cDNA clones in order to explore the entire coding sequence. Even if it is more time consuming, this method can detect, in some cases, a restriction fragment with an abnormal size, which can be used to perform a direct diagnosis in a female relative at risk. If no deletion has been identified, a duplication will be searched for by a careful examination of the Southern blots. Then, the gene dosage assessment will be performed using the same technique associated with a quantification using the PhosphorImager, or using real-time PCR (see Chapter 1). In some cases, a pulse field gel electrophoresis (see Chapter 1) might be required, or even FISH or a molecular combing.

Search for point mutations. The search for a point mutation is time consuming, difficult and expensive, and therefore it should only be performed secondarily. Even if many fast screening techniques have been described, the best method is the PTT due to the length of the gene and the frequency of truncating mutations (no-sense or stop). Finally, whatever the mutation, it is important that all the individuals suffering from muscular dystrophy in a family are studied to make sure that they all carry the same mutation. Indeed, families where some of the individuals carried different mutations (meaning the family mutation or a new mutation) have been reported. Considering that this type of event could affect 3 per cent of the familial forms, it is important to rule out this possibility before starting a family study or a prenatal diagnosis.

Family studies   The purpose of a family study is to determine the carrier status of females who are at risk. The strategy used will vary and depends on the accessibility of the proband and on the identification of the familial molecular defect. Usually, the family mutation will be searched for directly or, if it is unknown, an indirect family study will be performed using intragenic molecular markers (see Chapter 1).

Direct evaluation of the status of at-risk females. When the family mutation is known (deletion, duplication, point mutation identified in the proband or an obligatory female carrier) it will be directly searched for. The technique used will depend directly on the type of family mutation. The particular case of deletions removing a region containing one (or several) polymorphic marker(s) should be mentioned. In this case, it will be possible to observe a loss of heterozygosity in the carriers. In the same way, the presence of junction fragments will also be valuable in determining the status of at-risk females. Whatever the context of the diagnosis, it is important to use two different techniques performed in parallel and whose results should be identical. The purpose of this imperative is to compensate for the limitations of the methods used (with the Southern technique it is difficult to perform a gene dosage assessment, and with amplification it could be that null alleles exist or there is a defect in DNA repair or DNA replication after slipped-strand mispairing). Very often, a family study limited to the informative polymorphic markers flanking the mutation will be performed in parallel to the mutation search.

Indirect evaluation of the status of at-risk females. The dystrophin gene contains many polymorphic markers: RFLP, microsatellites and SNP. When it has not been possible to identify the family mutation, the status of the at-risk females will be evaluated through an indirect family study using markers. The main procedures of the strategies used are represented in Figure 2.21. The family study is especially difficult when it concerns the dystrophin gene because it is very unstable, and therefore hosts intragenic recombination events that will have to be taken into account.

   In the experience of the authors, a recombination is found in 35 per cent of tested families and it is quite frequent to observe two different events in the same family. Therefore, informative markers spread all over the gene and in flanking regions, distal and proximal, will have to be identified for each family, and even for each branch of a multiplex family. Finally, every analysis will have to be completed by a risk evaluation

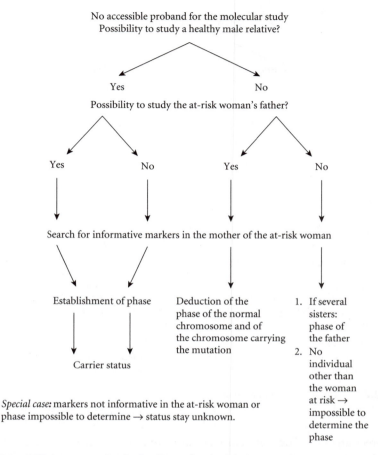

Figure 2.21   Different strategies for family studies in the absence of an accessible proband to perform a molecular study

for any studied woman. Different probabilities should be combined to integrate the a priori risk (according to the pedigree), the level of CK in the plasma, and the error risk in the molecular analysis. In this latter case, it will be the error risk due to a double recombination event that would not have been identified with the studied markers (most common case), or to an intragenic recombination (in case of a lack of information in the region of the gene).

Prenatal diagnosis

Molecular analysis. Prenatal diagnosis is usually performed using chorionic villus sampling performed at 10–13 weeks of amenorrhea, or using amniotic cells from amniotic fluid sampled either at 16 weeks of amenorrhea or after. The strategy used for the prenatal diagnosis will vary and, as for a family study, it will take into account the existence or not of the family mutation identification and therefore a direct

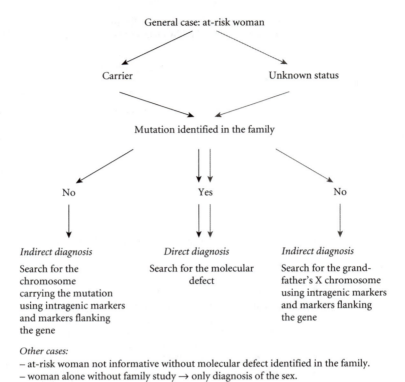

Figure 2.22   Strategy for prenatal diagnosis

or indirect analysis. The different steps are summarized in Figure 2.22. As for the evaluation of the carrier status of at-risk women, the analysis will be completed by a risk evaluation.

Dystrophin in utero *biopsy analysis.*   It has been possible to perform the dystrophin analysis using an *in utero* biopsy since 1993. This type of analysis is only suggested in special circumstances: deceased proband for whom no study has been made, fetus who has inherited his grand-mother's X-chromosome or an intragenic recombination visualized on the fetus haplotype, in a context where the family mutation has not been identified. In all of these circumstances, the mother's status is unknown, her total CK level is normal and the fetal risk is between 5 and 10 per cent. The biopsy is performed at about 32 weeks, a date where the biopsy is accessible and the fetus is viable. The biopsy is performed by trained obstetricians. Two samples are taken under sonogram and are analysed by Western blot (with two antibodies) and by immunofluorescence on tissue section. This test makes it possible to confirm or to formally exclude a diagnosis of Duchenne muscular dystrophy. It is only performed as an exception and raises a more general problem (obstetrical and ethical) of a very late pregnancy interruption if the answer turns out to be positive.

To conclude, even though 14 years have passed since the dystrophin gene was cloned, further studies are necessary to understand, at the molecular level, the mechanisms that make the gene so unstable and, at the protein level, the precise functions of dystrophin and of its different isoforms. In terms of diagnosis, the cloning of the gene and the identification of deletions and duplications makes the diagnosis much easier, both for the prenatal diagnosis and for the evaluation of the status of at-risk females. However, there are still cases for which the diagnosis is very difficult: death of the proband before any molecular study was performed, lack of family information making the detection of new mutations impossible, or sibships too small to do a family study. Even if these situations are less and less frequent, they are difficult to handle by the doctors and the biologists who are powerless and have to face ethical problems when asked for a prenatal diagnosis. These are cases where the only possibility is to offer a diagnosis of the sex of the fetus, with either a therapeutic abortion and a study of the dystrophin from the aborted fetus, or a study directly on fetal tissue at the end of the pregnancy, followed sometimes by a very late abortion. More efficient and cheaper molecular techniques are expected in order to solve these difficult situations.

2.4 Neurodegenerative diseases

2.4.1 Introduction

The neurodegenerative diseases constitute an important chapter in neurology. The most common are: Alzheimer's disease, Parkinson's disease and Charcot's disease (amyotrophic lateral sclerosis). These pathologies represent a public-health problem that gets worse with an aging population. Indeed, most of the time these diseases start in adulthood. At the pathophysiological level, a neuronal effect in a few neurons is observed which can expand during the evolution of the disease to many territories of the central nervous system (CNS).

Since case/control and twin studies have been performed, a genetic component has been suspected of playing a role in the determination of these diseases. However, it has not been possible to identify a simple transmission model (Mendelian inheritance, meaning monofactorial). Therefore, these diseases are called multifactorial diseases. Because it is very difficult to identify the genes responsible when the inheritance model is complex, many research teams have chosen to study the inherited neurodegenerative diseases for which a Mendelian transmission can be determined, allowing a strategy called 'positional cloning'. This is the case for Huntington's disease and other Mendelian neurodegenerative disorders like the autosomal dominant cerebellar ataxias (ADCA). This research has been very fruitful because on the one hand they have allowed the identification of a new type of mutation, and on the other hand they showed the link between the pathologic process and the presence of intranuclear inclusions, in particular degenerating neurons.

2.4.2 Polyglutamine neurodegenerative disorders

Dynamic mutations

During the last 15 years, a new class of mutations has been involved in diseases affecting the nervous system. These mutations are the result of DNA instability and, therefore, they are called dynamic mutations. They consist of trinucleotide repeats with a size of array normally stable from generation to generation, but not transmitted following the classical rules of genetics in patients. The number of repeats can vary from a few to several hundreds.

Two classes of dynamic mutations have been identified: the long expansions in non-coding regions (see Section 2.3, fragile X), associated with CGG, CTG or GAA trinucleotides, and the medium sized expansions in coding regions (exons) concerning the CAG, or GCG trinucleotides (Table 2.5). This section concerns the latter class. The

Table 2.5 Diseases associated with trinucleotide expansion in coding regions
Genes – IT15: important transcript 15; HD: Huntington's disease gene; PAPBD2: poly(A) binding protein 2; AR: Androgens' receptor; PPP2R2B: protein phosphatase PP2A regulatory subunit B, brain specific; TBP: TATA-binding protein
Transmission modes – XR: X-linked recessive; AD: autosomal dominant; AR: autosomal recessive

Disease	Gene	Heredity	Triplet	Site
Kennedy's syndrome (BSMA)	AR	XR	CAG	5' coding
Spinocerebellar ataxia 1 (SCA1)	SCA1	AD	CAG	5' coding
Spinocerebellar ataxia 2 (SCA2)	SCA2	AD	CAG	5' coding
SCA3/MJD (Machado–Joseph disease)	MJD1	AD	CAG	3' coding
Spinocerebellar ataxia 6 (SCA6)	CACNL1A4	AD	CAG	3' coding
Spinocerebellar ataxia 7 (SCA7)	SCA7	AD	CAG	5' coding
Huntington's disease (HD)	IT15 (HD)	AD	CAG	5' coding
Dentatorubro-pallidoluysian atrophy (DRPLA)	B37 (DRPLA)	AD	CAG	Central
Oculopharyngeal muscular dystrophy (OPMD)	PABP2	AD	GCG	5' coding
Ataxia associated with dementia	TBP	AD	CAG	Coding

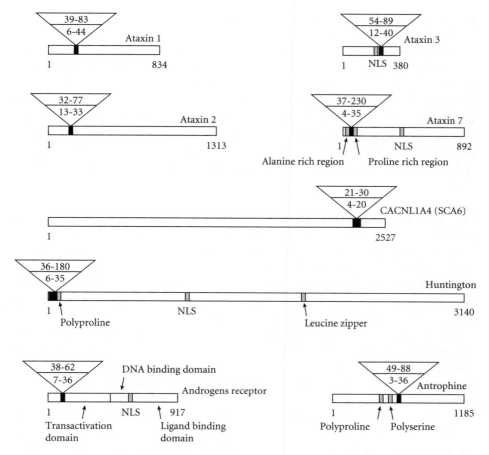

Glutamine repeat position in the proteins involved in disorders caused by the CAG tri-nucleotide expansion
The triangle indicates the position of the glutamine repeat. The bottom half of the triangle represents the
number of repeats in normal alleles while the top half indicates the size of the pathological alleles. The size
of the proteins in amino acids is indicated (NLS: nuclear localization signal).

Figure 2.23   Glutamine repeats

CAG or GCG trinucleotide repeat expansions, found in coding sequences, contain
less than 300 repeats that are translated as glutamine or alanine homopolymers in
the corresponding gene product (Figure 2.23).

The example of Huntington's disease

Clinical picture   Huntington's disease (HD) is the perfect example of an autosomal dominant neurodegenerative disorder. Its frequency is relatively high in Western
Europe – 5/10 000. It is characterized by the presence of involuntary movements associated with cognitive and psychiatric problems. The disease generally develops when

people are between 40 and 50 years old, more rarely in childhood or at a later age up to their 80s. The early symptoms are often anomalies of the ocular saccades – sharp, uncontrollable movements of the face and limbs – that turn into chorea (where the common name Huntington's chorea comes from). In 50 per cent of cases, psychiatric symptoms appear before the chorea. The juvenile form is characterized by rigidity in the first instance, discrete choreic movements and a fast progression of the disease leading to death in 8 to 10 years. The neuronal loss affects mostly the striato-efferent GABAergic neurons. The loss is more important in the caudate nucleus than in the putamen. Later, the cortex, the hypothalamus and the cerebellum are also affected, leading to a reduction of up to 20 per cent of the brain volume.

IT15 gene and responsible mutations   The gene responsible had been localized by linkage analysis on the short arm of the chromosome 4. After 10 years of genetic and physical mapping, the candidate region had been reduced to 2 megabases, entirely covered by yeast artificial chromosomes (YACs) and overlapping cosmids. A classic technique of positional cloning then allowed the identification of a 210 kb gene, important transcript 15 (IT15). The gene responsible for Huntington's disease contains 67 exons and seems to have a ubiquitous expression.

The most important discovery was the observation of a CAG trinucleotide expansion unstable in the first exon of the gene in patients affected with HD. On normal alleles, the size of the repeat varies from six to 35 CAG while pathologic chromosomes have between 36 and 180 CAG. The normal and the pathologic alleles are both transcribed and translated.

In the case of HD, the threshold is difficult to determine. Indeed, three groups of alleles should be considered depending on the number of repeats. The normal alleles present six to 35 repeats. The alleles with 41 to 180 CAG are always associated with the Huntington's disease. The intermediate alleles with 36 to 41 CAG are most of the time associated with the HD phenotype, but are also found in asymptomatic individuals at an advanced age. For this reason, these alleles are said to belong to the zone of reduced penetrance.

Genotypic diagnosis   This is based on the specific PCR amplification of the IT15 gene fragment containing the CAG expansion. Often, it is possible to distinguish pathologic alleles from normal alleles on an agarose gel by comparing the electrophoretic profile of healthy control individuals and affected individuals. The amplification products (amplicons) are then submitted to electrophoresis on a high-resolution polyacrylamide gel that can separate fragments with only a two base pairs difference. The co-migration of a size marker with the amplicons from the tested patients gives the number of CAG repeats for each of the two copies of the IT15 gene. The results are similar to those shown for fragile X (Figure 2.14). For the patients, two fragments (two alleles) are visible with a large size fragment corresponding to the pathologic allele. In healthy control individuals, the number of CAG repeats is often different for each allele of the IT15 gene but is always less than 35 repeats.

Negative correlation between the number of CAG repeats and the age of onset

In each disorder caused by a CAG trinucleotide expansion, a negative correlation between the age of onset and the number of CAG/glutamine repeats on the mutated allele has been observed. However, the large variation between the observed ages of onset for a given number of repeats makes this correlation impossible to use to predict the age of onset, especially in a presymptomatic diagnosis. For example, for a given number of triplets, the age at which patients develop the first signs of the disease can vary by as much as 40 years. Indeed, the number of repeats only accounts for 50 to 75 per cent of the variability of the age of onset, suggesting that other factors – genetic or environmental – are also involved.

The dynamics of unstable mutations

For a long time, clinicians have noticed in certain neurologic disorders (like HD), a phenomenon called anticipation: The age of onset decreases over successive generations and, often, some juvenile forms (before 20 years) will appear with paternal transmissions. The properties of the unstable mutations can now explain this observation.

There is indeed a replicative instability of the pathological allele through transmission, notably through paternal transmission. When the number of pathological alleles present in male sperm is compared with the number present in the lymphocytes, a much wider distribution of their size and a much higher average size can be seen in the sperm. Therefore, there is a high instability and a propensity for expansion of pathological alleles during spermatogenesis.

The correlation between the age of onset and the number of CAG repeats, together with the tendency for the repeat number to increase from generation to generation, explains the phenomenon of anticipation observed in HD and in other polyglutamine disorders. The number of CAG repeats also influences the clinical picture of the patient as shown for HD, BSMA and above all for ADCA. Finally, the large normal alleles (29–35 CAG repeats) – also called intermediate alleles – which are moderately and more rarely unstable, in particular during paternal transmission, could be at the origin of de novo mutations replacing, in the pool of pathologic alleles, those that have been eliminated by selection.

Medical applications of the molecular test and ethical implications

Diagnostic aids   The identification of mutations associated with a neurodegenerative disease allows for the confirmation of a diagnosis in a patient presenting some signs of the disease, thanks to a molecular test.

Presymptomatic diagnosis   The molecular test also allows the determination of the status of an asymptomatic individual with respect to the disease affecting a relative. This condition belongs to the application area of predictive diagnosis. It is often the situation for HD as well as for other polyglutamine diseases for which the age of onset is in adulthood. Presymptomatic testing raises ethical questions when the pathology is severe and when no preventive or curative treatment can be offered, as is the case for HD. Presymptomatic testing has been the object of serious thought at the international level. Some guidelines for presymptomatic testing have been determined: benefits, informed consent, autonomy of the individual, privacy, right not to know, equal access to tests. A protocol for taking charge of and following up the person requesting testing has been set up: it requires a multidisciplinary team that may include neurologists, genetic counsellors, social workers, psychiatrists and psychologists.

It has to be noted that if a person, still young, knows that he is a carrier for a pathological mutation, then it implies *ipso facto* that one of his parents is a carrier due to the autosomal dominant transmission which can generate a conflict between the right to know for one, and the right not to know for the other.

Prenatal diagnosis   This is rarely asked for in the first intention, for neurodegenerative diseases, notably for polyglutamine disorders. Indeed, it would mean indirectly performing a presymptomatic diagnosis of the parent at risk. Therefore, before planning a prenatal diagnosis that might be useless if the parent is not a carrier, a presymptomatic diagnosis is offered to the parent at risk. If someone knows that they are a carrier of a mutation for a neurodegenerative triplet disorder, this will often cause them to rethink their ideas of having children.

2.4.3   Pathophysiology of polyglutamine disorders

A common signature: the nuclear inclusions

The mechanisms by which the CAG/polyglutamine repeat expansions induce the selective neuronal death observed in these disorders are still unknown. However, it seems to be a gain of function; this means the acquisition by the corresponding protein of a novel and toxic function, in accordance with the dominant transmission mode of these diseases, and with the fact that the two alleles are expressed at the same rate in a given tissue. In all the diseases associated with a CAG trinucleotide expansion, there is a threshold in the number of repeats after which the pathology appears. This threshold varies depending on the locus and is 30 to 40 repeats, except in the case of SCA6 (20 repeats). How can proteins tolerating 20 to 35 glutamines become toxic for the cell after a threshold of 35? The physical and/or biological properties of these proteins must be altered.

All the immunohistochemical studies agreed on the fact that, at the opposite of the normal forms, the pathological forms of the proteins involved in those diseases have a nuclear localization within intranuclear inclusions in the affected tissues.

The discovery of an anatomopathological signature common to all these disorders is an important point from which many hypotheses can be made. Whether the intranuclear inclusions are the consequences or the direct cause of the disease, they constitute a good marker because their density is correlated with the size of the expansion. They are also found *post mortem*, on brain sections of the patients affected with HD. In the CNS, these inclusions have been found only in neurons, and especially in those located in the affected structure. However, there is no correlation for a given structure between the density of inclusions and the level of neuron loss.

Inclusion formation

In order to identify the mechanisms involved in these disorders, it is necessary to understand the steps leading to their formation. The first step of the pathologic process seems to be the conformation change of the protein due to the expansion. This phenomenon, at the origin of the gain of function, is ubiquitous because the pathological form of the protein can be detected in many structures, even the ones that are not affected, peripheral or central. This change of conformation is likely to induce the multimerization of these proteins but this phenomenon does not happen in any tissue. *In vitro* experiments showed that the polyglutamine expansions form insoluble filaments that could have an antiparallel β-sheet structure. Furthermore, the proteins with polyglutamine expansions constitute an *in vitro* substrate for the transglutaminases expressed in the nervous system. These enzymes could establish stable chemical bounds between polyglutamines of different mutated proteins.

Huntingtin, the product of the IT15 gene, should be transported in the nucleus to form aggregates and induce neurodegeneration. It is probable that the nucleus constitutes a primary site for the pathogenesis of the disorders with polyglutamine expansions. It is possible to speculate that once the polyglutamine proteins are present in the nucleus, they form aggregates through a mechanism depending on their concentration and the size of their polyglutamine domain.

The reason for the specificity of the defects, despite a ubiquitous expression of the mutated proteins, is still not understood. One of the hypotheses is that the specificity is due to the interaction of the mutated protein with partners specifically expressed in the affected tissues.

Inclusions in other mono- or multi-factorial neurodegenerative pathologies

The presence of intranuclear inclusions is one of the neuropathological signs common to polyglutamine diseases. The analysis of cell models and animal models demonstrate the link between the formation of the inclusions and the degenerative process of the neurons. What is the situation for common neurodegenerative diseases such as Alzheimer's (AD) and Parkinson's (PD) diseases?

AD has been known for a long time to be characterized by extracellular (senile plaques) and intracellular (neurofibrillary degenerations) deposits. The senile plaques are mostly composed of the β-amyloid peptide while the neurofibrillary degenerations contain the cytoskeleton protein Tau. The β-amyloid peptide is produced by cleavage of a transmembrane protein, the β-amyloid peptide precursor, which can be cleaved in different ways depending on the secretase involved in the process. The main β-amyloid peptide contains 40 amino acids (Aβ40) while Aβ42/43 is minor in physiological conditions. The current pathophysiological theory considers that the central event in AD is an overproduction of β-amyloid peptide. This hypothesis about the amyloid pathway is supported by the results of the study on the autosomal dominant forms of AD. In fact, the identified mutations in the gene encoding for the β-amyloid peptide precursor (APP) and in the genes encoding the presenilins affect the β-amyloid peptide metabolism by inducing an overproduction of its insoluble form, Aβ42/43. Aβ42/43 has the property to form aggregates more easily than the main form Aβ40, so can form amyloid fibres. The amyloid fibres then accumulate to form the senile plaques.

Parkinson's disease is another common neurodegenerative disorder whose prevalence reaches 2 per cent after the age of 70. The clinical manifestation is the trio of akinesia, rigidity and rest tremor, which respond favourably to a substitutive treatment with levodopa, at least at the onset of the disease. In fact, the disease is due to a neuronal loss preferentially affecting the neurons from the substantia nigra that produce dopamine. The lesions appear together with a histopathological marker, the Lewy body, eosinophile and ubiquitinylated cytoplasmic neuronal inclusions. The monogenic forms are considered as being rare but several genes responsible for Parkinson's disease have been identified: α−synuclein and the ubiquitin carboxy-terminal hydrolase L1 (UCH-L1) for the autosomal dominant forms, and Parkin for the autosomal recessive ones. Their identification could allow the understanding of some of the steps of the pathological process. Indeed, it is interesting to notice that α−synuclein constitutes a major component of the Lewy bodies present in idiopathic Parkinson's disease. In addition, the overexpression of α−synuclein in mice and in drosophila happens to be toxic for the neurons, especially the dopaminergic neurons.

Therefore, these elements together demonstrate the central place that both nuclear and cytoplasmic inclusions have in the pathophysiology of neurodegenerative diseases. It is also interesting to notice that the study of the monogenic forms, even though they are rare, is very informative. Indeed, it allows for the identification of common aspects concerning the neurodegenerative process. The family forms of AD and PD, like the neurodegenerative disease with polyglutamine expansions, are due to mutations that favour the formation of proteins with physiochemical characteristics different from that of the normal proteins (lower solubility, tendency to form aggregates, tendency to form structures poorly degradable by cellular catabolism, etc.), resulting in their accumulation within intra- or extracellular deposits that will have direct or indirect consequences on neuron survival. If the mechanisms leading to the formation of the inclusions are beginning to be elucidated, the initial steps of the degenerative process still need to be understood in order to identify new therapeutic targets.

2.5 References and Bibliography

2.5.1 References

Beutler, E. (1997). 'The significance of the 187 g (H63D) mutation in hemochromatosis.' *Blood Cells Mol. Dis., 23,* 135–145.

Blonden L. A. *et al.* (1991). '242 breakpoints in the 200-kb deletion-prone P20 region of the DMD gene are widely spread.' *Genomics, 10,* 631–639.

Brissot, P. (2000). 'Le diagnostic de l'hémochromatose à l'heure du test génétique.' *La Presse Médicale, 29,* 8.

Douabin, V. *et al.* (1999). 'Polymorphisms in the HFE gene.' *Hum. Hered., 49,* 21–26.

Eberhart, D. E. *et al.* (1996). 'The fragile X mental retardation protein is a ribonucleoprotein containing both nuclear localization and nuclear export signals.' *Hum. Mol. Genet., 5,* 1083–1091.

Eichler, E. E. *et al.* (1994). 'Length of uninterrupted CGG repeats determines instability in the FMR1 gene.' *Nature, Genet., 8,* 88–94.

Feder, J. N. *et al.* (1996). 'A novel MHC classI-like gene is mutated in patients with hereditary haemochromatosis.' *Nat. Genet., 13,* 399–408.

Fergelot, P., Lohyer, S. and Le Gall, J.-Y. (1998). 'HFE, une nouvelle molecule HLA de classe I, est impliquée dans le metabolisme du fer.' *Médecine/Sciences, 14,* 387–391.

Fu, Y. H. *et al.* (1991). 'Variation of the CGG repeat at the fragile X site results in genetic instability: resolution of the Sherman paradox.' *Cell, 67 ,* 1047–1058.

Jouanole, A. M. *et al.* (1996). 'Haemochromatosis and HLA-H.' *Nat. Genet., 14,* 251–252.

Lakich, D. *et al.* (1993). 'Inversions disruption of the factor VIII gene are a common cause of severe haemophilia A.' *Nat. Genet., 5,* 236–241.

Monaco, A. *et al.* (1988). 'An explanation for the phenotypic differences between patients bearing partial deletions of the DMD locus.' *Genomics, 2,* 90–95

Oudet, C. *et al.* (1993). 'Linkage disequilibrium between the fragile X mutation and two closely linked CA repeats suggests that fragile X chromosomes are derived from a small number of founder chromosomes.' *Am. J. Hum. Genet., 52,* 297–304.

Papanikolaou, G. *et al.* (2004). 'Mutations in HFE2 cause iron overload in chromosome 1q-linked juvenile hemochromatosis.' *Nat. Genet., 36,* 77–82.

Roberts, R. *et al.* (1993). 'Exon structure of the human dystrophin gene.' *Genomics, 16,* 536–538.

Roetto, A. *et al.* (1999). 'Juvenile hemochromatosis locus maps to chromosome 1q.' *Am. J. Hum. Genet., 64,* 1388–1393.

Roetto, A. *et al.* (2004). 'Screening hepcidin for mutations in juvenile hemochromatosis: identification of a new mutation (C70R).' *Blood, 103,* 2407–2409.

2.5.2 Bibliography

Bellon, G. (1995). 'Mucoviscidose.' *La Revue du praticien,* **45,** 2351–2360.

Chelly, J. and Kaplan, J.-C. (1988). 'La myopathies de Duchenne, du gène DMD à la dystrophine.' *Médecine/Sciences,* **4 ,** 141–150.

Claustres, M. *et al.* (2000). 'Spectrum of CFTR mutations in cystic fibrosis and in congenital absence of the vas deferens in France.' *Hum. Mutat.,* **16,** 143–156.

Cystic Fibrosis Genetic Analysis Consortium – Cystic Fibrosis Mutations Database : www.genet.sickkids.on.ca/cftr/

Dreyfus, B. *et al.* (1992). *L'Hématologie de Bernard Dreyfus* (1992), Médicin-Sciences Flammarion.

Durr, A., Lucking, C. and Brice, A. (2000). 'La maladie de Parkinson due aux mutations de la Parkine.' (Pour le groupe français de recherche génétique sur la maladie de Parkinson.) *Médecine/Sciences,* **16,** 1112–1115.

Emery, A. E. (1993). *Duchenne muscular dystrophy,* 2nd edn. Oxford Monographs on Medical Genetics Volume 24, Oxford Medical Publications.

Ghanem, N. *et al.* (1992). 'A comprehensive scanning method for rapid detection of β-globin gene mutations and polymorphisms.' *Hum. Mutat.,* **1,** 229–239.

Gilgenkrantz, H. (1993). *Dystrophine et myopathies. Pathologie moléculaire, expression et thérapie génique.* Monographie de l'Association française contre les myopathies (AFM).

Girodon-Boulandet, E., Cazeneuve, C. and Goossens, M. (2000). 'Screening practices for mutations in the CFTR gene ABCC7.' *Hum. Mutat.,* **15,** 135–149.

Gitschier, J. *et al.* (1984). 'Characterization of the human factor VIII gene.' *Nature,* **312,** 326–330.

Haemophilia B mutation database (http://europium.csc.mrc.ac.uk/usr/WWW/WebPages/main.dir/main.html)

HAMSTeRS: the haemophilia A mutation, structure, test and resource site (http://www.umds.ac.uk/molgen/haemBdatabase.html)

Hirst, M. C., Grawal, P. K. and Davies, K. (1994). 'Precursor arrays for triplet repeat expansion at the fragile X locus.' *Hum. Mol. Genet.,* **3,** 1553–1560.

Lebre, A. S. and Brice, A. (2001). 'Aspects génétiques et physiopathologiques des affections causées par une expansion de polyglutamine.' *La Lettre du neurologue,* **avril,** 34–37.

Longshore, J. W. and Tarleton, J. (1996). 'Dynamic mutations in human genes: a review of trinucleotide repeat diseases.' *J. Genet.,* **75,** 193–217.

Lunkes, A. and Mandel, J.-L (1997). 'Polyglutamines, nuclear inclusions and neurodegeneration.' *Nature, Med.,* **3,** 1201–1202.

Mandel, J.-L. (1991). 'Syndrome de l'X fragile: des mutations étonnamment ciblées et instables et un gene à la recherché d'une function.' *Med. Sci.,* **7,** 637–639.

Mandel, J.-L. (1997). 'Breaking the rule of three.' *Nature,* **386,** 767–769.

Mornet, E. and Simon-Bouy, B. (1996). 'Biologie moléculaire du syndrome de l'X fragile: données récentes et applications diagnostiques.' *Archives de Pédiatrie,* **3,** 814–821.

Myers, R. H., Marans, K. S. and MacDonald, M. E. (1998). 'Huntington's disease.' In *Genetic Instabilities and Hereditary Neurological Diseases* (Wells, R. D. and Warren, S. T., eds). San Diego, Academic Press, pp. 301–323.

Navarro, J. and Bellon, G. (2001). *La mucoviscidose.* Editions Espaces 34, 2nd ed.

OMIM (Online Mendelian Inheritance in man) http://www.ncbi.nlm.nih.gov/entrez/query.fcgi?db=OMIM

Orphanet (database of rare diseases) : http://orphanet.infobiogen.fr/

Roberst, R. G., Gardner, R. J. and Bobrow, M. (1994). 'Searching for 1 in 2,400,00: a review of dystrophin gene mutations.' *Hum. Mutat.*, **4**, 1–11.

Rosenstein, B. J. and Cutting, G. R. (1998). 'The diagnosis of cystic fibrosis: a consensus statement.' Cystic Fibrosis Foundation Consensus Panel. *J. Pediatr.*, **132**, 589–95.

Stamatoyannopoulos, G. *et al.* (1994). *The Molecular Basis of Blood Diseases.* Philadelphia, W. B. Saunders Company.

The globin gene server (http://globin.cse.psu.edu), Containing a large number of links to exhaustive data banks with information on quantitative and qualitative globin anomalies.

Welsh, M. J. *et al.* (1995). 'Cystic fibrosis'. In *The Metabolic and Molecular Bases of Inherited Disease*, (eds Scriver, C. R. *et al.*). New York, McGraw-Hill Health Professions, pp. 3799–3876.

Welsh, M. and Smith, A. (1996). 'La mucoviscidose.' *Pour la Sciences*, **220**, 66–74.

Yoshitake, S. *et al.* (1985) 'Nucleotide sequence of the gene for human factor IX (antihemophilic factor B).' *Biochemistry*, **24**, 3736–3750.

3 Molecular diagnosis in oncology

Dominique Stoppa-Lyonnet, *Institut Curie, Paris*

3.1 General introduction

Cancer is a genetic disease of the cell. Transformation of a normal cell into a cancer cell is linked with the accumulation of chromosome alterations progressively leading to a cell in an undifferentiated state, with local and metastatic proliferation and diffusion capabilities. Tumourigenesis is based on alterations that give a selective advantage to the cell compared with others; this selection process is analogous to the one occurring during the Darwinian evolution, proceeding by successive random changes and selections. The number of alterations necessary for a normal cell to be transformed varies from one type of tumour to another one but is likely to be small, estimated as being between four and seven (Renan, 1993). These alterations, or mutations, happen spontaneously because of DNA replication errors during cell division, or are induced by mutagenic agents. In most cases, they only appear in tumour cells or cells about to be tumourous: these are somatic mutations.

The cytogenetic and molecular genetic developments that have taken place during the last 20 years have seriously modified our knowledge of the mechanisms of oncogenesis, allowing identification of recurrent anomalies in all cancers or in site-specific cancers. The aim of the cytogenetic and molecular study of cancers concerns both the diagnosis and prognosis, therapy and prevention. Even if the description of the genetic anomalies leading to a tumour is far from finished, the first clinical applications are appearing. In this chapter, the main cellular pathways affected during the formation of a tumour and the different types of genetic alterations leading to it will be schematically described. A few examples will illustrate how genetic studies can have an effect on the diagnosis, prognosis and therapy. Genetic predisposition

Diagnostic techniques in genetics   Edited by Jean-Louis Serre
© 2006 John Wiley & Sons, Ltd

to cancer will then be discussed, and the molecular diagnoses already available and how they are put to use will be presented.

3.2   Cellular pathways targeted by the tumour process

Even though the number of genetic alterations is extremely large, preventing an exhaustive follow-up of the literature in this field, it is possible to have a broad idea if they are pooled depending on the selective advantages they give to the cell. Hanahan and Weinberg (2000) recently defined six classes of selective advantages:

- independence of the cell from the effects of growth factors

- loss of negative control of cell proliferation

- avoidance of apoptosis

- unlimited cell division capabilities

- angiogenesis

- local diffusion and metastasis.

The acquisition of these selective advantages seems necessary for a tumour to develop, but the genes involved in conferring a particular advantage can vary from one type of cancer to another. For example, it has been shown in colon cancers that the tumours carrying a mutation in the APC gene do not have a mutation in the gene encoding β-catenine and vice versa, while the products of both of these two genes are involved in the negative control of cell proliferation.

1. Cell proliferation depends on extracellular growth factors that bind trans-membrane cell receptors activating a cascade of proteins leading to cell prolif-eration (mitogenic signal transduction). To become independent of extracellular growth factors, the cell can: (a) synthesize its own growth factors and create an autocrine stimulation loop (synthesis of tumour growth factor, TGFβ, and glioblastoma); (b) have a permanently activated growth factor receptor (activat-ing mutation in the *Ret* gene leading to a permanent tyrosine kinase activity in the thyroid medullar cancers); (c) gain a permanent activity of the protein upstream of the growth factors receptors (Ras protein activated in 50 per cent of colon cancers).

2. Cell proliferation also depends on anti-proliferative signals that block the cell in the G1 phase of the cell cycle – the post mitotic state which corresponds to a cell differentiation state – or that remove the cell from the cell cycle, leaving it in a quiescent G0 state. One of the best-known proteins involved in the G1-S transi-tion control (where S is the DNA synthesis phase) is RB (retinoblastoma).

The de-phosphorylated RB protein sequestrates the protein E2F, a transcription factor controlling the expression of a large number of proteins directly involved in the G1/S transition and in DNA synthesis. Conversely, phosphorylation of RB frees E2F allowing the transcription of key proteins of the cell cycle. RB phosphorylation itself depends on kinases and kinase inhibitors like CDK4 and P16 respectively. Some of these kinase inhibitors themselves depend on extracellular factors like TGFβ. Another pathway leading to the cell being in the G1 phase is the APC-β catenin pathway that is very often inactivated in colon cancer as we saw before. Therefore, it is easy to understand the tumourigenic effect of a homozygous loss of function mutation in the RB gene.

3. Apoptosis is a cell death driven by an endogenous programme and started when a cell becomes useless or dangerous for the organism. Apoptosis occurs during development or when a cell accumulates a number of genetic alterations that it is not able to repair or that are critical for cell survival. Two types of proteins are involved in apoptosis: the proteins leading the cell toward apoptosis and that are directed by p53, and the protein effectors represented by the caspases. p53 inactivation is one of the most frequent processes in the tumourigenic process to escape apoptosis being found in more than 50 per cent of cancers.

4. Cell division is limited – after a certain number of divisions, the cell enters into senescence. One of the mechanisms leading to senescence is the shortening of the telomeres by 50 to 100 bases pair at each cell cycle, a phenomenon that is absent in 85 to 90 per cent of the cancers. During the tumourigenic process the telomeres are protected by a telomerase that adds the missing nucleotides.

5. Neovascularization or angiogenesis is 'indispensable' to the development of tumour cells that need oxygen and nutriments. Angiogenesis is driven by a delicate equilibrium between angiogenic factors (vascular endothelial growth factor (VEGF) for example) and anti-angiogenic factors (thrombospondin 1, interferon β, etc.). The regulation of these factors is affected during the tumourigenic process. Considering the important number of identified angiogenic and anti-angiogenic factors, there are a huge number of possible mechanisms for tumourigenic angiogenesis.

6. Tissue homeostasis is maintained thanks to inter-cellular interactions between cell–cell adhesion molecule (CAM) proteins like E-cadherin. The local and metastatic diffusion abilities of a cell can result from a loss of expression of the CAM proteins, from the expression of proteases in the extra-cellular matrix, or from an adaptation of the cell to its new environment thanks to the expression of new integrins (cellular proteins that bind the extracellular matrix). Nevertheless, mechanisms leading to the modification of the expression of these different proteins are still not well understood.

3.3   Types of genetic alteration leading to cancer

3.3.1   Introduction

Among the mutations involved in oncogenesis, we distinguish the gain of function mutations (also called activating mutations) that correspond to an increase of activity or to the gain of a new activity, from the loss of function mutations (see the introduction to Chapter 2). Due to the functional effect of these two types of mutation in tumour genesis, molecular biologists defined the affected genes as oncogenes in the first case because the activation of the gene is part of the tumourous process, or as tumour suppressors in the second case because the loss of function of the gene is associated with the tumourous transformation, meaning that the physiological function of these genes is the suppression or the protection of the cell against tumourigenesis.

3.3.2   Activating mutations

The activating mutations lead by definition to the activation of an oncogene. Only one of the two alleles needs to be activated to have an oncogenic effect (dominant effect, see introduction to Chapter 2). In general, these are missense point mutations (amino acid substitution). The *Ras* genes, whose codons 12, 13, 59 and 61 are frequently mutated, constitute one of the most-studied gene families in oncology. These mutations keep Ras proteins in an activated state that has been shown to play a role in mitotic signal transduction. A second type of activating mutation causes an increase in copy number for a gene, for example, the amplification of the N-*myc* gene in the neuroblastoma. A third example is the activation of a gene due to the translocation of its coding sequence next to the promoter of another gene. In Burkitt's lymphoma, the c-*myc* gene is activated by the translocation of its coding sequence (chromosome 8) next to the promoter of one of the immunoglobulin genes (most of the time on chromosome 14, and sometimes on chromosome 2 or 22) that are constitutively transcribed in B lymphocytes, leading to an abnormal and high expression of the c-*myc* protein. A fourth example, an oncogene can be fused to another gene to create a chimeric gene with a higher oncogenic activity. In chronic myeloid leukaemia(CML), the ABL gene, normally located on chromosome 9, is fused with the BCR gene on chromosome 22 to create the BCR-ABL hybrid gene that has a constitutive tyrosine kinase activity.

3.3.3   Inactivating mutations

Inactivating mutations lead to the inactivation of the corresponding protein. In general, an inactivating mutation of one allele is associated with the inactivation of the second allele (which is necessary to have a complete loss of function of the corresponding protein, see introduction to Chapter 2). Inactivating mutations consist of point mutations leading to a stop codon, small deletions or insertions causing a frameshift or creating a stop codon, or missense mutations causing a loss of function of the protein. The large diversity of inactivating mutations compared with the activating mutations can easily be explained by the fact that there are multiple types of mutations and multiple sites whose mutation can cause a loss of function. The inactivation of a gene can be the result of its complete deletion or the deletion of its entire region. Chromosome deletions are intensively studied, and one of the most common methods is to search for the loss of an allele in a tumour, called loss of heterozygosity. It is an indirect way to detect a deletion performed by a comparative analysis of the alleles found in normal tissue and those found in a tumour (see Chapter 4).

Finally, inactivation of a gene can result from an epigenetic mechanism like methylation of the promoting region or the sequestration of a viral protein: the RB protein and the E7 protein of the papilloma virus (strain 16 and 18) in uterine cancers.

3.4   Alteration origins: the role of the repair genes

Mutations happening during a human's lifetime can be due to mutagenic agents like tobacco, X-rays, UV-rays but can also be the results of DNA replication errors during the S phase of the cell cycle (see Table 3.1). These random alterations on the genome can affect a gene coding for a key process in oncogenesis, making it easier for the cell to enter one of the selection processes mentioned in the introduction. Statistical studies have shown that, in the absence of anomalies in the DNA repair mechanisms, the probability for these mutations to accumulate in one single cell or its daughter cells was low compared with the frequency of cancers in the population. Therefore, it has been suggested that defects in the DNA repair process are a prerequisite for the mutations to accumulate. During the last few years, the identification of mutations in DNA repair genes in several colon and breast cancers linked with a genetic pre-disposition made this hypothesis more likely. Kinzler and Vogelstein (1997) introduced the notion of *caretaker* genes to designate the DNA repair genes as opposed to the genes involved in one of the six key steps of oncogenesis mentioned above, which are called *gatekeepers*.

3.5   Benefits of molecular studies to patient healthcare

In many haemato-oncology pathologies, non-random chromosome anomalies appear during the transformation. These chromosome anomalies are termed primary and can be specific for a given haemopathy and therefore help with the diagnosis. Other anomalies can appear during evolution of the disease and these additional

Table 3.1　Rare repair anomalies

Predisposition	Main tumour risks and manifestations	Course of action	Estimated frequency* of the carriers in the general population (G) and among the cancer cases (C)	Genetic study(ies)
Ataxia telangectasia	Haemopathy Neurological problems	Hematological surveillance, treatment adaptation, genetic counselling	(G) 1/40 000 to 1/300 000	ATM, hMRE11 Karyotype
Fanconi syndrome	Haemopathies	Treatment adaptation, genetic counselling	(G) 1/350 000	Crosslinking agents sensibility tests, Six identified genes
Bloom syndrome	Multiple cancers	Non-specific surveillance, genetic counselling	(G) 1/1 000 000	BLM, sister chromatid exchange tests
Xeroderma Pigmentosum	Skin cancers	Dermatologic surveillance, genetic counselling	(G) 1/500 000 to 1/1 000 000	UV sensibility tests, Eight genes
Werner syndrome	Premature aging, cancers	Non-specific surveillance, genetic counselling	(G) 1/300 000 to 1/1 000 000	WRN

* These estimates are given as an indication.

anomalies are termed secondary and are the sign of a progressing disease. Among the acute myeloid leukaemias, all chromosome anomalies do not have the same meaning for the diagnosis, some give a good prognosis, others a bad prognosis, so their characterization is therefore important in choosing a good therapy.

3.5.1   Chronic myeloid leukaemia (CML)

CML is a myeloproliferative syndrome with hyperplasia of the granulocytes. After several years, the disease progresses faster and becomes acute leukaemia. The CML is the first neoplasic pathology that has been described which is associated with an acquired genetic anomaly, an exchange of a part of the long arms of chromosomes 9 and 22 called translocation t(9;22)(q34;q11). This translocation is found in 90 per cent of the CML and is called CML Ph$^+$, the derived chromosome 22 being the Philadelphia chromosome (because it was described in this town in 1960) (see Figure 3.1(a)).

The BCR-ABL chimeric fused gene (Figure 3.1(b)) is leukaemogenic because the protein has a constitutive tyrosine kinase activity, activating the transduction pathways and interfering with many cellular processes like proliferation, adhesion or apoptosis. Tyrosine kinase inhibitors have been developed and Imatinib mesylate, a specific inhibitor for the ABL tyrosine kinase, gives exceptionally good results in the therapeutic tests (Mauro *et al.*, 2002). The BCR-ABL kinase activity inhibition leads to a transcriptional modulation of the genes involved in the cell cycle control, in adhesion and in cytoskeleton organization, causing the apoptosis of the Ph$^+$ cells. The BCR-ABL fusion can be detected in the CML Ph$^-$ by molecular cytogenesis using fluorescent probes located at the level of the breakpoints.

The CML diagnosis can be performed using regular cytogenesis, molecular cytogenesis or molecular biology using RT-PCR. To follow the evolution of the disease during treatment, it is possible to evaluate the cytogenetic response depending on the percentage of Ph+ cells left. In regular cytogenetics about 30 mitoses are studied, and in molecular cytogenesis, between 200 and 500 nuclei are tested. It is also possible to evaluate the molecular response using quantitative RT-PCR techniques like TaqMan®quantitative PCR. These molecular biology techniques have the advantage of being more sensitive and can detect one BCR-ABL+ cell out of a million. It is therefore possible to obtain a complete cytogenetic answer (0 per cent of Ph$^+$ cell) with still a certain level of detection of the transcription product BCR-ABL. To follow the response to a treatment, the molecular biology is more sensitive. However, the cytogenetics can detect secondary chromosome anomalies that are the first sign of transformation into an acute leukaemia like the appearance of a chromosome 17q (Figure 3.1(a)). It is therefore justified to follow the evolution both with cytogenetics and molecular biology, both techniques giving complementary pieces of information.

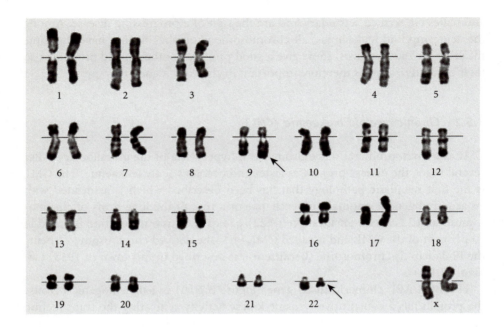

BCR

e1	e2	e3		e12	e13	e14	e15	e16		e23
				b1	b2	b3	b4	b5		

    ↑   ↑   ↑

   P190   P210

ABL

a1	a2	a3		a11

    ↑

P190 BCR-ABL

e1	a2	a3		a11	e1-a2

P210 BCR-ABL

e1	e2	e3		e12	e13	a2	a3		a11	b2-a2
				b1	b2					

e1	e2	e3		e12	e13	e14	a2	a3		a11	b3-a2
				b1	b2	b3					

3.5.2 Acute myeloid leukaemia (AML)

The AML are classified using the FAB (French American British) classification, from AML0 to AML7 depending on the maturation of the myeloid blasts. The presence of certain chromosome anomalies specific for a type of AML helps diagnosis. The translocation between the long arms of chromosomes 8 and 21 t(8;21)(q22;q22) is specific of certain AML2 and leads to the fusions of the genes AML1-ETO that can be detected using RT–PCR. The AML3 are characterized in 94 per cent of cases by the t(15;17)(q22;q21), fusing the gene coding for the retinoic acid receptor α RAR, on chromosome 17 to the gene PML on 15. However, in 4 per cent of cases, the fusion PML-RARα is infracytogenetic, meaning there is no visible chromosome re-arrangement. The gene rearrangement can only be detected by FISH using specific probes or by RT–PCR using primers located on both sides of the fusion gene. Lastly, 2 per cent of the AML3 have another translocation, implicating RARα on the 17 but PML on the 15. It is indispensable to detect the t(15;17) or the PML-RARα fusion because the only AML3 that respond to the retinoic acid treatment leading to a normal maturation of the blasts are those having this rearrangement. The value of certain chromosome anomalies for prognosis is well known for AML. The transloca-tions t(8;21), t(15;17) and the chromosome 16 inversion are good prognostics while conversely, complex karyotypes (more than three anomalies), monosomies of either the entire 5 and 7 chromosomes or just the long arms, the translocation t(6;9) and t(9;22) and anomalies of the long arm of chromosome 3 are bad prognostics. The choice of therapies depends on which of the two prognostic groups the anomaly belongs to. All the other chromosome anomalies, as well as a normal karyotype, are interpreted as intermediate risks for the patient. However, the karyotype is normal in 30 to 50 per cent of AML cases, and it is particularly interesting for these cases to find some genetic anomalies in order to distinguish disease types having a differ-ent prognosis. Currently, several genes for AML are studied using the Southern blot technique or sequencing: FLT3, MLL, CEBPα, N-ras and C-kit. FLT3 is a tyrosine

---

(a) Philadelphia chromosome: t(9;22) translocation in chronic myeloid leukaemia generates a fusion between the gene c.abl in chromosome 9 and the bcr region in chromosome 22. The presence of an isochromosome 17q (two times the long arm, secondary classic anomaly for CML indicating the acutization) can be observed (photograph from Jérôme Couturier, I. Curie).

(b) C-abelson fusion protein: this exchange of chromosome material fused the gene BCR to the gene ABL on chromosome 22 and ABL to BCR on chromosome 9. In the 10 per cent of CML cases without the t(9;22), called Ph⁻, the gene rearrangements can be detected one time out of two using RT–PCR with primers located on both sides of the created fused gene. At the level of ABL, the breakpoint is most of the time between exons 1a and 1b, giving the junction a2. At the level of BCR, the breakpoint is most of the time in a 5kb region called MBCR (stands for major breakpoint cluster region) between exons b2 and b3 or between exons b3 and b4, giving the transcription products b2a2 or b3a2 respectively and coding for a 210kDa protein (p210$^{bcr-abl}$). More rarely, the break is upstream, between exons e1 and e2, giving a transcription product with the first abl exon e1a2 coding for a 190kDa protein (p190$^{bcr-abl}$).

Figure 3.1

kinase receptor. There are two types of mutation for this receptor: internal tandem duplication or point mutation. These mutations lead to the receptor activation with dimerization and autophosphorylation. FLT3 duplication is found in 23 per cent of AML cases, mostly in those with a normal karyotype, and a point mutation in 6 per cent of the cases. It seems that FLT3 duplications are associated with a bad prognostic. Several inhibitors of FLT3 tyrosine kinase have been developed and are currently being studied in humans. A molecular biological study of the genetic anomalies in AML cases is essential in addition to a cytogenetic study.

3.5.3 Burkitt's lymphoma

Burkitt's lymphoma is a proliferation of one hyperbasophile lymphoid B precursor poorly differentiated with a peculiar morphology. It is rare in the occident but endemic in Africa, in the region infested with malaria. The Epstein–Barr virus (EBV) is found at the level of the tumour in 95 per cent of the endemic forms and in only 15 per cent of the sporadic forms. A polyclonal expansion of B lymphocytes can be caused for example by malaria. Then, chromosome translocations will put the c-myc oncogene in 8q24 under the control of the strong promoter of one of the immunoglobulin genes, most of the time IgH in 14q32, IgK in 2p12 or IgL in 22q11, and its over expression will favour proliferation of the cell. Lastly, cells with a cytogenetic rearrangement will be immortalized by the EBV.

The karyotype therefore shows a t(8;14)(q24;q32) or a variant translocation t(2;8)(p12;q24) or t(8;22)(q24;q11). The rearrangement of c-Myc next to one of the immunoglobulin genes can be shown with molecular biology by RT–PCR using specific primers located in both parts of the genes.

3.5.4 Anatomopathologic diagnosis, therapy and prognosis

Anatomopathologic diagnosis is not always easy, especially in the case of round cell sarcomas that cover a heterogeneous set of tumours. Their diagnosis has been greatly improved by the identification of specific translocations, detected using a cytogenetic approach or a molecular approach with, in the latter case, the detection of the fusion transcription products by RT–PCR. The diagnosis of Ewing's sarcomas, malignant melanomas of soft tissues, desmoplastic tumours, myxoid chondrosarcomas, alveolar rhabdomyosarcomas, synovialosarcomas can be confirmed by only one amplification step using primers specific for each translocation (Delattre and Dauphinot, 1998).

Some other therapeutic applications for these inhibitors are being developed, especially for some rare digestive tumours called gastro-intestinal stromal tumours (GIST), that almost always have a defect in the c-kit oncogene that also has a physiological tyrosine kinase activity. Another therapeutic application can be cited, with a monoclonal antibody, trastuzumab or Herceptine®. Certain breast cancers, with bad

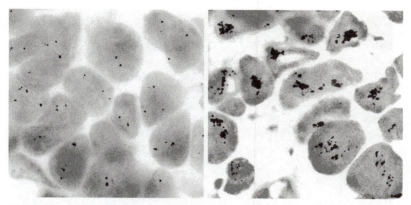

(a) Breast cancer without HER2 amplification (b) Breast cancer with HER2 amplification

Revealed by hybridization using a specific DNA probe (FISH) (photograph by Jérôme Couturier).

Figure 3.2 HER2 amplification and breast cancer

prognoses, are characterized by the amplification of the HER2 growth factor, a trans-membrane receptor that also has tyrosine kinase activity (Figure 3.2). Herceptine® recognizes and specifically inactivates HER2. Its efficiency for the treatment of these cancers has been shown recently (Hortobagyi, 2001).

The study of the genetic characteristics of cancer can also help with their prognosis. One of the most obvious and most used factors for the clinical aspect is the amplification of the transcription factor N-myc in neuroblastoma. The amplification, sometimes resulting in a very large number of copies (several hundreds), is visible by FISH on a biopsy (Figure 3.3). The amplification of N-myc is a bad prognostic factor, independently of the extension state of the lesion. Its detection leads to a more intensive therapy.

3.6 Genetic predisposition to cancers

3.6.1 Introduction

The transformation of a cell is linked with the accumulation of a certain quota of mutations that most of the time appear only in the tumour cells or in the cells about to become tumour cells (see introduction to Chapter 3), and are somatic mutations. These somatic mutations are different from the constitutive or germ mutations, present in all cells of the organism, and especially in the germ cells. The quota of alterations necessary for cell transformation is reached faster in some individuals who carry: (a) a mutation that participates in tumour formation and that is present constitutively, mutation in a 'gatekeeper gene', or (b) a constitutive mutation in a

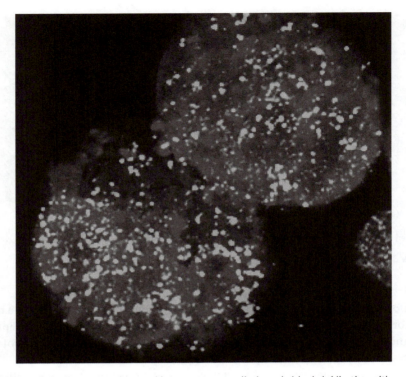

Amplification of the N-myc gene in neuroblastoma tumour cells (revealed by hybridization with a specific DNA probe (FISH)). Note the amount and the dispersion of the copies in the genome (photograph by Jérôme Couturier I.Curie).

Figure 3.3   Neuroblastoma and N-myc amplification

'caretaker gene' involved in the genome stability or in the detoxification of mutagenic agents, leading to a large number of somatic mutations.

An individual carrying one or other of these genetic constitutive alterations has a higher risk of developing a cancer than anyone else in the general population, and can also potentially transmit this predisposition to their children, the alteration affecting the germ cells. The importance of the risk and the tissues at risk depend on the function of the affected gene in the tumour formation process and on the tissue it is expressed in (Thomas, 1995).

Predispositions linked to genetic factors outside of the cell have not been considered here. However, they are likely to be frequent, could concern a large number of cases in the population, could occur via interactions with environmental factors and could be associated with low tumour risks. A variation in the hepatic enzymes involved in the detoxification of mutagenic agents, or the sensitivity of individual immune responses to oncogenic viruses like some papillomaviruses, are good examples.

3.6.2   Genetic predispositions to cancers identified so far

In 15 years, more than 30 genes giving a predisposition to cancer have been identified. From rare familial childhood cancers (retinoblastoma, nephroblastoma, Table 3.2) to common familial cancers (colon cancer, breast cancer, Table 3.3), and including the tumour predisposing syndromes whose associated symptoms are in the forefront of cases presenting at clinics (Von Recklinghausen's disease or type I neurofibromatosis, ataxia-telangiectasia), most of the severe genetic predispositions to cancer have been studied and the genes responsible have been identified.

This research was inspired and stimulated by the fact that the identification of the genes responsible could give the same number of pathways to understand cancer formation, and also because they could provide a basis for genetic tests to find out whether or not an individual with a family history is predisposed to cancer, and to adapt the healthcare provided accordingly.

The genes predisposing to cancer belong to several of the gene categories already mentioned: oncogenes, tumour suppressor genes, repair genes (caretaker genes). However, only four oncogenes have been identified so far as being at the origin of a predisposition: *Ret, Met, C-kit, CDK4* (Tables 3.3 and 3.4).

Since the mutations of tumour suppressor genes and caretakers are inactivating mutations, there is (as previously mentioned) a considerable diversity of mutations that can only be detected using a complete screening of the suspected gene(s). Because of these difficulties with the investigation, the study is first done in the individual who has the most chance of being predisposed, taking into account their personal history (having had a cancer) and their family history. The first individual being tested is, by definition, the 'proband'. When a mutation has been identified in the proband, then a test can be offered to the relatives. Even though the testing of the relatives is easy, based on the mutation previously identified, the test in the proband is complex because it requires a complete study of one or several genes. Lastly, a negative result (no mutation identified) often gives little information because it does not rule out an underlying predisposition. It is, however, reassuring for the relatives.

The tests offered at the moment in medical practice are those meeting the three criteria:

- risk of tumour based on epidemiologic data published and validated by independent studies

- taking care of the patient and their relatives with a defined risk – that is to say a risk defined by experts or by consensus from a multidisciplinary group

- indications from the genetic tests performed on the proband.

Cancers are extremely frequent so every cancer family history does not signify an underlying predisposition and, because of the limited significance of a negative result

Table 3.2 Predisposition to rare cancers discovered in childhood

Predisposition	Main tumour risks	Course of action	Estimated frequency* of the carriers in the general population (G) and among the cancer cases (C)	Identified gene(s)
Retinoblastoma	Retinoblastoma, Sarcoma	Monthly postnatal eye exam or prenatal diagnosis	(G) 1/40 000 (C) 1/3	RB
Li–Fraumeni	Sarcoma, breast, CNS surrenal	Clinical exam, imaging guided by the clinical exam	(G) 1/30 000 (C) 1/100 childhood sarcomas	P53
Nephroblastoma, Nephropathies, WAGR syndrome, BW	Kidney, Hepatoblastoma, cortico-surrenaloma	Surveillance by renal sonogram (cancer and nephropathy), hepatic sonogram depending on the context, genetic counselling	(G) 1/100 000 (C) 1/20	WT1 11p15 (uniparental disomy) KIP2p57
Rhabdoid tumours	Brain or kidney localization	Limited, genetic counselling		SNIF/INI1

* These estimates are given as an indication.

Table 3.3 Predisposition to cancers without associated predisposing diseases, generally appearing in adulthood

Syndrome's name	Main tumour sites	Course of action	Estimated frequency* of the carriers in the general population (G) and among the cancer cases (C)	Identified gene(s)
HNPCC** syndrome	Colon, endometrium	Colonoscopy twice a year after 25 years old, endometrium surveillance from 30 years old	(G) 1/500 (C) 1/20	hMLH1, hMSH2 (hMSH6)
Breast cancer	Breast, ovary	Mammography from 30 years old, ovariectomy between 35 and 50 years old, possibility of mastectomy	(G) 1/500 (C) 1/30	BRCA1, BRCA2
Melanoma	Skin melanoma	Skin exam every 6 months and photo every 12 months, excision of doubtful lesions	(G) 1/1 500 (C) 1/20	CDKN2A, CDK4
Papillary kidney cancer	Kidney cancer	Renal scan	(G) 1/5000 to 1/10 000 (C) 1/20	c-MET
Gastro-intestinal stromal tumours	Stomach T, Small intestine and colon T,	Sonogram, abdominal scan, endoscopy	—	C-kit
Stomach cancer (other than HNPCC)	Stomach	Gastric fibroscopy	(G) 1/10 000 to 1/20 000 (C) 1/100 to 1/200	CDH1 (E-cadherin)

* These estimates are given as an indication.
** HNPCC stands for hereditary non-polyposis colorectal cancer.

Table 3.4   Familial multiple endocrinopathies

Predisposition	Main tumour sites	Course of action	Estimated frequency* of the carriers in the general population (G) and among the cancer cases (C)	Identified gene(s)
Multiple endocrine neoplasia type 1	Parathyroid hyperplasia and thyroid, pancreas, small intestine tumours	Guide to the surgery, serum test, target organs imagery	(G) 1/30 000 to 1/40 000	MEN1
Carney's syndrome	Cardiac myxoma, pituitary adenoma, testicular tumour	Cardiac sonogram, pituitary MRI	(G) 1/60 000 to 1/80 000	PRKAR1A
Multiple endocrine neoplasia type 2	Medullary carcinoma, thyroid, pheochromocytomas	Calcitonin dosage, complete thyroidectomy	(G) 1/30 000 to 1/40 000 (C) 1/10 to 1/20	RET

* These estimates are given as an indication.

in the proband, it is at present necessary to limit the conclusions from a genetic test to the most likely situations (Tables 3.1 to 3.5.). In the future, when a negative result in a proband could rule out the presence of a predisposition factor and when the laboratories' capacities are greater, then the conclusions from tests could only be expanded.

3.7   Genetic tests for cancer predisposition

3.7.1   Introduction

The way to proceed for prediction in oncology is based on the principle that the identification of a tumour risk in an individual should help in taking better care of them, and that a negative result should be reassuring. Even if these perspectives are truly a common point, the reality of genetic testing in 2006 is in fact more complex, and is very variable from one predisposition situation to another. The ethical problems generated by oncogenetics are becoming more important as the efficiency of treatments is limited and negative tests do not rule out the presence of an underlying predisposition. These problems will be illustrated using three different cancer predisposition situations.

Table 3.5 Hamartomatosis with dominant tumour expression

Predisposition	Main tumour sites	Course of action	Estimated frequency* of the carriers in the general population (G) and among the cancer cases (C)	Identified gene(s)
APC	Colon	Colonoscopy from 10 years old	(G) 1/8000 (C) 1/100	APC
VHL	Kidney, brain tumour, surrenal gland	CNS MRI, abdominal scan, urine test, eye exam, starting from childhood	(G) 1/40 000	VHL
NF2	Peripheral nervous system tumour	Audiogram, CNS MRI	(G) 1/30 000	NF2
Peutz–Jeghers	Cervix, colon, pancreas, breast	Smear test, colonoscopy, barium meal test, gastroscopy	(G) 1/50 000 to 1/100 000	LKB1
Gorlin syndrome	Basal cell carcinoma, medulloblastoma	Skin surveillance and orthodontic surveillance	(G) 1/50 000 to 1/100 000	PTCH
Cowden, Banayan–Zonana	Breast, thyroid	Mammography, thyroid sonogram	(G) 1/50 000 to 1/100 000	PTEN

* These estimates are given as an indication.

3.7.2   Help taking care of at-risk individuals: multiple endocrine neoplasia type 2

Multiple endocrine neoplasia type 2 (MEN2) are the best examples of the interest of cancer predisposition testing. The MEN2 (Table 3.4) are characterized by a major risk of medullary thyroid carcinoma (MTC), close to 90 per cent. The risk exists from childhood and is associated with a risk of parathyroid hyperplasia and pheochromocytomas (Figure 3.4). The potential severity of the disease raises the option, before the age of 10, of a preventive thyroidectomy associated with a substitutive hormonal treatment in predisposed children. The *RET* gene, whose alteration is responsible for MEN2, encodes for a transmembrane protein with a tyrosine kinase activity. Germline mutations are not frequent and are mostly localized in the transmembrane domain or in the kinase domain of the protein. Also, the relationship between the position of the mutations in the gene and the importance of the risk is known. For example, the risk of having a pheochromocytoma is seven times higher with a mutation in codon 634 compared with a mutation in codon 618. Finally, an estimate of 5 per cent of sporadic MTC, with no family history are, however, associated with a constitutive mutation in the *RET* gene. In most cases, these are neo-mutations appearing in the gametes of one of the two parents. It is now a common practice to offer a study of the *RET* gene for each case, even sporadic, of MTC. Therefore, the relative simplicity of the molecular analysis due to the absence of genetic heterogeneity of the disease (predisposition associated with a single gene: *RET*), and the low diversity of the mutations in this *RET* gene, the correlations

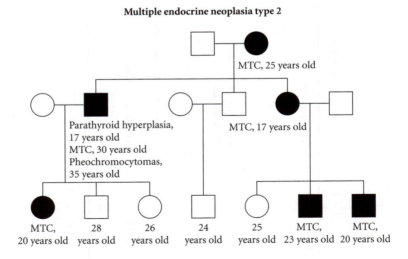

MTC stands for medullary thyroid carcinoma

Figure 3.4   Pedigree of a MEN2 family

genotype–phenotype and the tangible possibilities for prevention, made the predisposition tests for MTC a model situation in oncogenetics. The performance of the genetic tests, even in the sporadic cases of MCT, confirms their use in medical practice.

3.7.3   Some genetic tests are of limited interest: search for a constitutive mutation in the TP53 gene

In contrast to the previous situation, the search for mutations in the *TP53* gene involved in Li–Fraumeni syndrome (LFS) (Table 3.2) is far from being part of common medical practice. Indeed, the importance of the risks, the diversity of the tumour sites, the types of cancer for most of which an early diagnosis does not modify the chance of recovery, make taking care of patients carrying a constitutive mutation of the *TP53* gene extremely difficult. Besides a mammography screening initiated at an early age, before 30, and an annual clinical exam, there are no other preventive possibilities at the moment (Frébourg *et al.*, 2001). Nevertheless, it is important to give an answer to those families who want to know the origin of their terrible family history, and to the individuals who need to know whether or not they carry the *RP53* mutation identified in their father or mother. For example, in the author's department, there was the case of a couple who lost one son to glioblastoma at the age of six, another son to sarcoma at 10, and where the mother had an osteocarcoma followed by a breast cancer (Figure 3.5). Even though they refused a genetic consultation several times after their oldest son had been diagnosed, they wanted a genetic study after the diagnosis of the second son. During the informed consent process, they expressed clearly that they needed to understand, to know and to be able to point to the origin of their drama in order to rebuild and assimilate the situation,

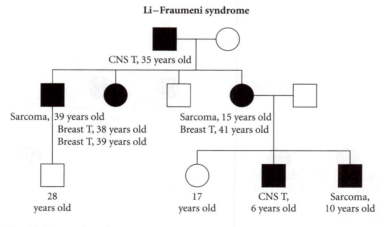

Li–Fraumeni syndrome

CNS T, 35 years old

Sarcoma, 39 years old
Breast T, 38 years old
Breast T, 39 years old

Sarcoma, 15 years old
Breast T, 41 years old

28 years old

17 years old

CNS T, 6 years old

Sarcoma, 10 years old

Figure 3.5   Li–Fraumeni pedigree

which had been unbearable up to then. The request level for genetic testing in such a situation will probably stay low. However, those who want it should still be able to access a genetic test.

3.7.4   Molecular genetic testing for breast cancer predisposition through the search for mutations in the BRCA genes

The predispositions to breast cancers (Table 3.3) combine all of the difficulties in genetic diagnosis and appear as an intermediate situation between the two extremes described above (Figure 3.6). Indeed, not all breast cancer family histories indicate a genetic predisposition, some familial clustering could be fortuitous and reflect the frequency of the disease in the population. For example, three sisters affected with breast cancer between the age of 60 and 69 have a genetic predisposition probability of about 20 per cent.

Two predisposition genes were identified in 1994 and 1995: *BRCA1* and *BRCA2* (BRCA stands for breast cancer). However, the alterations are in general different from one family to another, and more than 1500 different mutations in *BRCA1* and *BRCA2* have been identified.

They are dispersed on very long coding sequences, making family studies complex and, most of all, time consuming (see the problems with cystic fibrosis, Chapters 2 and 7). This is the reason why the strategies known as scanning and, in particular, the denaturing high-performance liquid chromatography (DHPLC) system (Chapter 1) as shown in Figure 3.7, are preferred. The sensitivity of the classical strategies to detect *BRCA1* and *BRCA2* mutations is estimated at 70 per cent. Finally, *BRCA1* and *BRCA2* are only involved in 65 per cent of hereditary cases, suggesting that one or several more genes still have to be identified. Therefore, the incomplete sensitivity

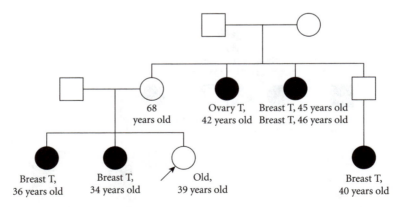

Figure 3.6   Breast cancer pedigree

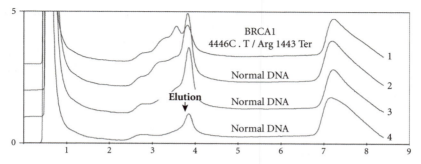

PCR fragments elution profile on a denaturing high-performance liquid chromatography (DHPLC) system (see Chapter 1). The profiles 2, 3 and 4 are identical and characteristic of a standard sequence; profile 1 is characteristic of the presence of a heteroduplex obtained in a carrier of a BRCA1 mutation (substitution of a cytosine with a thymine leading to a stop codon instead of an arginine codon). Ordinate: optical density; abscissa: minutes.

Figure 3.7   Detection of a BRCA1 mutation using DHPLC

of the analysis methods and the presence of other breast cancer predisposing genes limit the significance of a negative result after a first family study, because a genetic predisposition cannot be ruled out. However, finally, when after a long search a woman happens to be carrying a predisposition genetic factor giving her an 80 per cent risk of developing a breast cancer before the age of 80, and a 20 to 40 per cent risk for an ovary cancer, what type of preventive care can be offered to her? At the moment, she has two main options: frequent surveillance examinations initiated at an early age or prophylactic surgery. Even though mammography surveillance starting at the age of 30 in women at high risk has not been formally shown to lower the mortality by breast cancer, it is reasonable to extrapolate the results from previous screenings to this group of patients and to make assumptions about its efficiency. In contrast, to choose prophylactic surgery means to limit the maximum risk of cancer but at the cost of a mutilating and irreversible intervention. It is indeed reasonable to hope that in a few years time, chemoprevention of breast cancer could be available. A first promising study showed that Tamoxifen®, a molecule blocking oestrogen receptors, decreases by half the risk of breast cancer in at-risk women.

3.8   Conclusions and perspectives

The development during the last 25 years of cytogenetic and molecular genetic techniques has made it possible to comprehend the main mechanisms of tumoural transformation. A new pharmacopoeia is appearing thanks to the understanding of these mechanisms. Its leaders are the tyrosine kinase inhibitors. The current development

Table 3.6 Between dysmorphology, neurogenetics, dermatogenetics and oncogenetics:
the fringe cases

Predisposition	Main tumour sites	Course of action	Estimated frequency* of the carriers in the general population (G) and among the cancer cases (C)	Identified gene(s)
NF1	Skin signs, central and peripheral NS tumour	Clinical, CNS MRI	(G) 1/3000	NF1
Tuberous sclerosis (Bourneville disease)	Mental retardation, epilepsy, CNS tumour, kidney tumour, skin lesions	Clinical, kidney surveillance, neurogenetic counselling	(G) 1/10 000 to 1/15 000	TSC1, TSC2

* These estimates are given as an indication.

of high turnover techniques making possible the description of all the genetic and
protein alterations of a tumour should ensure the definition of a personalized treatment, adapted to each tumour.

The domain of cancer predisposition is a more recent domain of knowledge that
started in 1986 with the identification of the predisposition gene to retinoblastoma,
the *RB1* gene. More than 24 cancer predispositions are now part of the medical
practice. However, it is important to keep in mind that the cancer predispositions discovered up to now are the tip of the iceberg. They are the most obvious predisposition situations because they are the rare familial cancers (retinoblastoma), the severe forms of frequent familial cancers (colon cancer, breast cancer)
or finally, diseases associated with a tumoural risk (ataxia-telangectasia or risk of
haemopathy). The submerged part of the iceberg, very important by definition,
still has to be discovered and concerns the predisposition associated with a low tumoural risk and whose origin is multifactorial, combining environmental and genetic
factors.

This research, like all research done on the living, constitutes an important economical issue. The opposition of several cancer research centres and French hospitals,
raised at the European patent office concerning a patent granted to Myriad Genetics in relation to cancer predisposition tests based on the *BRCA1* gene's sequence,
illustrates this issue. The objective of the opponents was to challenge any monopoly
on genetic testing that would have a blocking effect instead of contributing to the
research progress (Gad *et al.*, 2001) and, therefore, to public health, which is the
main purpose of the patents.

3.9 References

Delattre, O. and Dauphinot, L. (1998). 'La famille des tumeurs d'Ewing.' *Med. Ther. Ped.*, **1**, 139–143.

Frébourg, T. *et al.* (2001). 'Le syndrome de Li et Fraumeni: mise au point, données nouvelles et recommendations pour la prise en charge.' *Bull. Cancer*, **88**, 581–587.

Gad, S. *et al.* (2001). '"Peignage d'ADN" et grands réarrangements du gène BRCA1 ou comment dénoncer le monopole de Myriad Genetics sur les tests de predisposition.' *Médecine/Sciences*, **17**, 1072–1075.

Hanahan, D. and Weinberg, R. A. (2000). 'The hallmarks of cancer.' *Cell*, **100**, 57–70.

Hortobagyi, G. N. (2001). 'Overview of treatment results with trastuzumab (Herceptin) in metastatic breast cancer.' *Semin. Oncol.*, **6** (suppl. 18), 43–47.

Kinzler, K. W. and Vogelstein, B. (1997). 'Cancer-susceptibility genes. Gatekeepers and caretakers.' *Nature*, **386**, 761–763.

Mauro, M. J. *et al.* (2002). 'STI571: a paradigm of new agents for cancer therapeutics.' *J. Clin. Oncol.*, **20**, 325–334.

Renan, M. J. (1993). 'How many mutations are required for tumorigenesis? Implications from human cancer data.' *Mol. Carcinog.*, **7**, 139–146.

Thomas, G. (1995). 'Dix ans de recherche sur les predispositions génétiques au développement de tumeurs.' *Médecine/Sciences*, **11**, 336–348.

4 Applications of molecular biology to cytogenetics

Étienne Mornet and Brigitte Simon-Bouy,
Université de Versailles

4.1 Introduction

Many chromosome anomalies in humans have been described (de Grouchy and Turleau, 1982; Gardner and Sutherland, 1996). Even if the karyotype (see Box 4.1) remains the only way to have an exhaustive view of all the chromosomes and of any eventual chromosome anomalies, the molecular biological methods are a useful complement in many cases, in particular for fast identification of anomalies in the number of chromosomes, of microrearrangements, and uniparental disomies.

4.2 Molecular diagnosis of anomalies in the number of chromosomes

4.2.1 Introduction

Anomalies in the numbers of chromosomes 13, 18, 21, X and Y constitute 80.5 per cent of the chromosome anomalies detected by karyotype analysis during pregnancy, which corresponds to one case out of 286 births (Boué, 1989). For a long time, the main criteria that have been used to justify a prenatal diagnosis are a mother being over 38 years old (the risk of having a child with a trisomy 21 is about one out 150), biological indicators (maternal serum markers) and ultrasound warning signs, which are now well defined. Currently, the diagnosis rests on the fetal karyotype performed after withdrawing a sample of amniotic fluid (amniocentesis) and cell culture. The

Diagnostic techniques in genetics   Edited by Jean-Louis Serre
© 2006 John Wiley & Sons, Ltd

Box 4.1   The Karyotype

The setting up of the karyotype allows a global visualization of the 46 chromosomes in 23 pairs, 22 pairs of autosomes and one pair of sex chromosomes. It can be performed using any nucleated cells (the cells most often used are blood lymphocytes and amniocytes) that are cultivated in an appropriate medium to obtain metaphasic mitoses, the stage at which the chromosomes are most condensed and therefore the best stage for analysis (with optical microscopy at a magnification of 1000). The chromosomes are then at a stage of the cell cycle where they have been duplicated without undergoing a mitotic division. They therefore appear in a linear form of two sister chromatids linked by the centromere (each chromatid containing exactly the same genetic material).

The homogenous colouring of the chromosomes is obtained with the help of a dye, the Giemsa (standard colouration). This allows the description for a chromosome: a short arm called p and a long arm called q, separated by the centromere. Certain techniques (controlled denaturation) cause the appearance of dark and light bands specific to each chromosome. Essentially, the R bands (obtained after heat treatment of the chromosomes) and the G bands (the chromosomes are treated with trypsin) can be distinguished. The topography of the G bands is reciprocal to that of the R bands. DNA fractionation on a caesium gradient has shown that the distribution of bases CG and AT is far from uniform on the chromosomes: the R bands are richer in CG and the G bands are richer in AT bases. The chromosome markings from the bands allow the chromosomes to be identified and classified by pairs according to a regularly updated international nomenclature (the nomenclature in question is ISCN 1995) and to possibly determine break points. It is typical to work with a level of resolution of 400 to 550 bands per haploid genome; it is possible to arrive at a larger number of bands (850) by using particular culture techniques, which will stretch the chromosomes (called a high-resolution karyotype). One of the large advances in cytogenetics in the 1990s was its combination with molecular hybridization allowing the use of DNA probes coupled to a fluorochrome directly on the slides of chromosome preparations. These probes could be specific to a locus, a band, an arm or an entire chromosome allowing visualization of the interesting region under the microscope using UV light, and its possible deletion or displacement in the case of a chromosome translocation.

amniocentesis can be performed from the 14th week of amenorrhea and the result is obtained within 10 to 15 days because of the requirement to culture the amniotic cells. It is also possible to perform the karyotype using chorionic villus biopsy as early as the 11th week of amenorrhea but, again, the cells need to be cultured for the result to be reliable.

The delay between sampling and the result of the karyotype is long and creates anxiety in the parents, while the ultrasound warning signs are visible earlier and are more evocative. For these reasons, it is interesting to have a rapid method for the prenatal diagnosis of anomalies in the number of chromosomes whose result could be given before the karyotype, which should be done anyway because it is the only test showing all chromosome anomalies.

Apart from this, some situations exist where the karyotype cannot be performed due to a failure in cell culture (*in utero* fetal death, sampling performed late in the pregnancy). For these cases, alternative methods have been developed using *in situ* hybridization techniques or PCR amplification of DNA sequences called microsatellites (or VNTRs, see Chapter 1).

4.2.2   *Diagnosis using fluorescence* in situ *hybridization*

Fluorescence *in situ* hybridization (FISH) consists of using molecular probes specific for chromosomes 13, 18, 21, X and Y, labelled with a fluorochrome, that are hybridized directly on the uncultured amniotic cells or the cells from the chorionic villus mounted on slides. A microscope with UV light is used for viewing the fluorescent spots in interphase nuclei of uncultured cells. In a normal fetus, two spots can be seen for the two chromosome 13s, two spots for the 18, two spots for the 21, one spot for the X chromosome and one spot for the Y chromosome (in the case of a boy) and two spots for the two X chromosomes (in the case of a girl). In the case of a trisomy, three spots will be seen instead of two. The use of different colour fluorochomes allows detection of several chromosomes in one nucleus. Large prospective studies (on about 150 000 karyotypes) have shown that this method is reliable considering that a large number of nuclei are analysed for each fetus.

4.2.3   *Molecular biology diagnosis*

Microsatellite markers

Microsatellite markers (see Chapter 1) are multiallelic genomic DNA sequences that consist of a motif – most of the time a di-, tri- or tetra- nucleotide motif – repeated in tandem a variable number of times. If for a given microsatellite motif, the number of repetitions is highly variable, the number of distinct alleles at this locus is very high. Therefore, most of the individuals in a population have a high probability to be heterozygous, and any two individuals have a high probability not to have any allele in common.

Microsatellites are frequent and evenly distributed throughout the genome so the probability of finding two individuals with the same genotype for all these loci is almost zero, meaning that all of the genotypes of one individual together constitute a unique signature or genetic 'identity card' (see Chapter 6).

These are the markers used for the molecular diagnosis of abnormalities in the number of chromosomes. In the case of trisomy 21, it is possible to select highly informative markers that allow the diagnosis of trisomy 21. For example, the D21S11 locus contains a complex repeated tetranucleotide sequence of the type of TCTA or TCTG that exists with 14 different allelic forms in the Caucasian population. The trisomy 21 is detected when there are three distinct alleles, or two distinct alleles with one being represented twice as much as the other. The latter case requires performing the PCR using quantitative conditions in order to obtain a gene dosage (see later). The only situation where the diagnosis is impossible is the case where a fetus carries three chromosomes with the same allele for the marker, which is then called non-informative. It is then necessary to use other microsatellite markers on the same chromosome. The polymorphism of these markers is such that it is rare to have several non-informative markers in the same family.

The different possible genotypes for a trisomic foetus

The informativity of a microsatellite marker depends essentially on its heterozygosity; for a fetus to carry at least two distinct alleles, it is necessary to have at least two distinct alleles segregating in the parents.

The informativity of a microsatellite marker also depends, when it is close to the centromere (see later), on the nature of the non-disjunction. Retrospective studies have shown that in two thirds of the cases the non-disjunction occurs at the first meiotic division, and in the other third of cases at the second division (Antonarakis, 1991; Muller *et al.*, 2000).

(a) *Non-disjunction at the first meiotic division without crossing over.* The gamete will be a carrier of two distinct parental chromosomes and two distinct alleles of the studied marker if the parent is heterozygous for the marker.

(b) *Non-disjunction at the second meiotic division without crossing over.* The gamete will be a carrier of a single parental chromosome duplicated and of two identical alleles for all markers.

(c) *Non-disjunction at the first meiotic division with crossing over.* As a function of chromatid segregation at meiosis II this produces:
- one time out of two (case no. 2) disomic gametes carrying all of the two chromosomes from a single parent (in an altered form or not)
- one time out of two (case no. 1) carrying one entire chromosome from one parent and one altered chromosome so just a fragment (centromeric or telomeric) from the other parent.

(d) *Non-disjunction at the second meiotic division with crossing over at meiosis I.* The gamete will carry two identical alleles for all markers located between the centromere and site of the crossing over (prereduction), and two different alleles for the markers located after the site and for which the parent was heterozygote (postreduction)

Figure 4.1   Schematic representation of chromosome non-disjunction

- In the case of a non-disjunction at the first meiotic division (meiosis 1), in the absence of crossing over (Figure 4.1(a)), the fetus inherits three distinct parental chromosomes – two from the parent where the non-disjunction occurred, and one from the other parent. The three chromosome 21s can carry one, two or three distinct alleles, the probabilities of the different configurations is linked to the

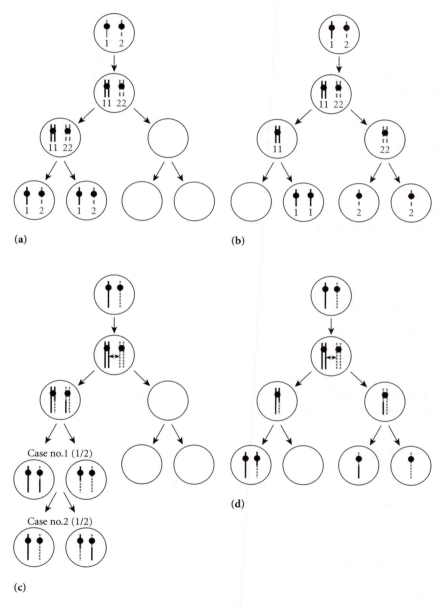

Figure 4.1 (*Continued*)

heterozygote status of the parents, and so to the heterozygosity of the marker. If the marker is very polymorphic, the probability of observing three alleles becomes important.

- In the case of a non-disjunction at meiosis 2, in the absence of crossing over (Figure 4.1(b)), the two chromosomes from the disomic gamete are identical, and the fetus will have at the most two distinct alleles, one of which will be in a double dose.

- In the case that a crossing over occurs, the situation is more complicated and requires distinguishing between the markers centromeric and telomeric vis-à-vis the CO.

In fact, in the case of a CO, the chromatid pairs are not altered before the CO and carry identical alleles (paternal for one of the pairs, maternal for the other), and they are altered after the CO and each pair of chromatids then carries a paternal allele on one chromatid and a maternal allele on the other (Figure 4.1(c) or (d), third line: homogenous with solid or dotted lines before the CO, heterogeneous that is to say altered) after the CO).

   If the non-disjunction occurs at meiosis 1 (Figure 4.1(c)), the fetus inherits a paternal and a maternal allele, systematically for markers centromeric to the CO, but only two times out of four for the markers telomeric to the CO. In contrast, if the non-disjunction occurs at meiosis 2 (Figure 4.1(d)), the fetus inherits a paternal and a maternal allele, systematically for the markers telomeric to the CO, and always two identical alleles, paternal or maternal, for the markers centromeric to the CO.

   In these conditions the informativity of a marker depends firstly on its heterozygosity (different maternal and paternal alleles), then its distance from the centromere as a function of the non-disjunction event; if this event occurs in meiosis 1 the centromeric markers will be more informative, if the event occurs in meiosis 2 the telomeric markers will this time be more informative. Figure 4.2 shows an example of the possible genotypes at a microsatellite locus for a trisomic fetus.

Molecular diagnosis

This depends on PCR amplification of the microsatellite markers and the use of electrophoresis to distinguish the alleles present.

- If the qualitative analysis allows three types of amplimers to be distinguished on the basis of their size (Figure 4.2, case 1), the molecular diagnosis of trisomy is qualitatively confirmed.

- If the qualitative analysis only allows one type of amplimer to be distinguished (Figure 4.2, case 4), molecular diagnosis is impossible because the result would be

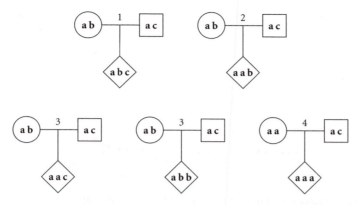

The fetus could inherit two distinct alleles from its mother and a third distinct allele from its father (case no. 1), two distinct alleles from its mother and a paternal allele identical to one of the two maternal alleles (case no. 2), a single maternal allele present in two copies and a distinct paternal allele (case no. 3) and, lastly, a single maternal allele present in two copies and an identical paternal allele (case no. 4). Only the last case does not allow the detection of a trisomy 21. Its frequency depends on the heterozygosity of the marker in the population, and also on the distance between the marker and the centromere (see Figure 4.1).

Figure 4.2   Segregation of the alleles of a microsatellite in the case of a maternal origin trisomy 21 (95 per cent of cases)

the same in the case of a non-trisomic fetus who was homozygotic for the marker allele; the studied marker is not informative and it is advisable to study another marker.

- If the analysis reveals the presence of two types of amplimers distinguished by their sizes (Figure 4.2, cases 2 and 3), the distinction between a trisomic fetus carrying two distinct alleles and a non-trisomic fetus also carrying two alleles rests on the number of copies of each of the alleles. It is therefore necessary to use 'quantitative PCR'.

PCR is only quantitative when in the part of the amplification curve where the number of copies amplified is proportional to the initial number of copies. It is therefore necessary to limit the number of PCR cycles so that the process does not reach the plateau of the curve where the number of copies obtained no longer depends on the initial number of copies (Figure 4.3). Quantitative PCR is delicate in the sense that there is only a narrow window between the minimum number of cycles required to obtain sufficient signal, and the maximum number of cycles so as to keep within the exponential part of the curve. This window varies as a function of the initial number of copies, that is to say the nature of the sample and the term of the patient. In practice, the number of PCR cycles to be performed is of the order of 20 to 30; Figure 4.4 shows an example of quantitative PCR in the case of marker D21S11 used in the molecular diagnosis of trisomy 21.

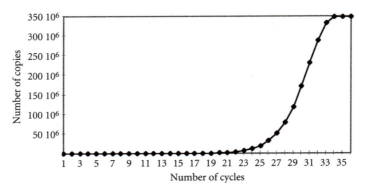

In this example, amplification is exponential in the first 28 cycles of PCR with an amplification rate close to 2^n where n is the number of cycles. After that, change in the composition of the reaction medium (exhaustion of free nucleotides and primers, high concentration of product, diminution of the enzymatic activity of DNA polymerase) leads to a diminution of the rate that will progressively approach one (plateau). It is also clear that it is necessary to perform enough cycles to obtain a detectable PCR product. The dosage window appears here between cycles 23 and 33.

Figure 4.3   Representation of the PCR curve

4.3   Chromosomal microdeletions

4.3.1   Introduction

A chromosomal deletion is the loss of part of a chromosome, leading to a partial monosomy for the chromosome fragment in question; they appear most frequently as *de novo* mutations (the karyotypes of the two parents are normal) and are most often internal (not reaching the most extreme parts of the chromosome). The first case – described in 1963 and which has become the most classic – is that of a deletion responsible for 'cat cry syndrome' (Lejeune *et al.*, 1963). This is caused by the loss of

→

The fetal genomic DNA is extracted from 2 ml of amniotic fluid taken during an amniocentesis and not cultivated. The genotype of the fetus for the microsatellite marker D21S11 is determined by PCR following an electrophoretic migration on an automatic sequencer like the ABI PRISM 310. The PCR products are tagged using a fluorochrome detectable by its fluorescence when excited by a laser beam. The co-migration of a standard-sized marker (peak heights corresponding to fragments from 100 to 250 bp) allows determination of the size of the PCR products and therefore identification of the alleles. In this example, the parental genotypes for the marker are indicated, but are not necessary for the diagnosis.

(a) Control fetus. The fetus has inherited the allele $(TCTN)_{31}$ from his mother as a single copy and allele $(TCTN)_{28}$ from his father. Quantitative PCR shows that the number of copies of each of these two alleles is the same. It can also be seen that the detection threshold of the signals emitted by the PCR products occurs at around 25 PCR cycles.

(b) Trisomic fetus having inherited the two maternal alleles $(TCTN)_{27}$ and $(TCTN)_{32b}$ and the paternal allele $(TCTN)_{28}$, detected after 27 PCR cycles.

(c) Trisomic fetus having inherited two copies of the maternal allele $(TCTN)_{26}$ and a copy of the paternal allele $(TCTN)_{31b}$, detected after 27 PCR cycles.

Figure 4.4   Molecular diagnostic of trisomy 21 using the microsatellite marker D21S11

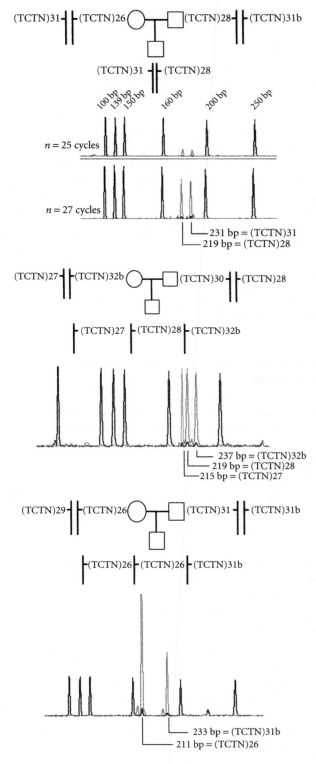

Figure 4.4 (*Continued*)

a variable sized fragment of the short arm of chromosome 5 (partial monosomy 5) responsible in newborns for a group of symptoms associated with a very characteristic high plaintive cry after which the syndrome has been named. Other deletions have also been described as being associated with syndromes of malformations with mental retardation, the example Wolf–Hirschhorn's syndrome and the deletion of a of the short arm of 4, or Williams' syndrome and a deletion situated under the centromere of 7 (which amongst other things involves a deletion of the elastin gene).

Depending on its size, the deletion will be more or less visible by classical cytogenic methods. When deletions cannot be seen using standard karyotyping they are called microdeletions. Actually, more and more clinical syndromes are described as being associated with microdeletions. Among the more classical examples is DiGeorge's syndrome (microdeletion on the long arm of chromosome 17 carrying in particular the LIS1 gene which plays a role in brain morphogenesis). These microdeletions occur most frequently *de novo* (that is to say that they do not exist in the parents of the affected child), but in rare cases they can be transmitted with variable phenotypes. It is very important to have methods that allow diagnosis of the syndrome in the child and genetic counselling for the family.

4.3.2 Mechanisms for the generation of microdeletions and microduplications

The observation that deletions and microdeletions associated with clinical syndromes occur most often *de novo*, but always in the same genomic regions, provides evidence that certain regions are more susceptible to deletions. Several hypotheses have been proposed for their generation, in particular the existence of repetitive sequences that favour bad meiotic pairings that lead to deletions/duplications, that is to say a duplication on the homologous chromosome to that which carries the deletion. Therefore, there exist some loci for which a syndrome is described for a microdeletion and another syndrome for the corresponding duplication; for example, Smith–Magénis syndrome (microdeletion at the level of the short arm of chromosome 17) and the 'dup(17)(p11.2) syndrome', which is the corresponding duplication with a clinically different phenotype, or Charcot–Marie–Tooth disease type 1A (duplication at the level of the long arm of chromosome 17) and hereditary neuropathy with a sensitivity to pressure palsies (the corresponding deletion). It must be noted that there are more syndromes associated with deletions than with duplications.

In certain cases the microdeletion is associated with a reciprocal translocation. A reciprocal translocation is an exchange of chromosome segments between two non-homologous chromosomes. All chromosomes can be affected and all break points are possible. When the translocation occurs, there can be a loss of chromosomal material (deletions or microdeletions) at the break point that can possibly be detected by the same methods previously described with the proviso that microsatellites or corresponding probes are available.

Finally, the regions located in the sub-telomeric part of chromosomes are, without doubt, the zones most sensitive to microdeletions because of their strongly repetitive structure. In this case as well, to detect the deletions, it is necessary to have microsatellites or specific probes (probes called 'all telomeres' are available commercially, but are not used much because of their high cost).

4.3.3   Methods to detect chromosomal microdeletions

Fluorescence in situ hybridization (FISH)

Molecular cytogenetics uses hybridization with fluorescent probes (the FISH method, see Chapter 1), which can be direct or indirect. In the indirect method, a DNA probe specific for the deleted region is attached to a protein (for example biotin, see Chapter 1) that has a strong cytochemical affinity for a molecule attached to a fluorochrome (for example avidin). Another system frequently used is that of digoxigenin (the DNA probe is marked with digoxigenin and the fluorochrome with an antibody against digoxigenin). The direct method uses a DNA probe specific for the deleted region already coupled with a fluorochrome. In all cases the marked loci are immediately visible under a microscope equipped with an ultraviolet light. When there is no microdeletion in the tested region, two fluorescent spots are seen on the two homologous chromosomes, whereas in the case of a microdeletion only one spot on the normal chromosome will be seen.

Using combinations of probes marked with different fluorochromes simultaneously, several chromosome regions can be hybridized in the same mitosis and possibly visualized with different colours. Typically, a probe specific for the microdeletion is used with a probe (called the control) for another region of the chromosome – for example centromeric or telomeric – which allows localization of the mitotic chromosome. In the case of a microdeletion, only the control spot is seen (Figure 4.5), while the normal chromosome has two spots (for the control locus and the studied locus). A certain number of probes specific for known chromosome microdeletions are commercially available.

As with microsatellites, the use of fluorescent probes is specific for a locus and so is subordinate to the clinical data. A simple and effective method to detect a priori all chromosome microdeletions does not currently exist.

Use of microsatellite markers

Microsatellite markers have frequently repeated sequences and are distributed throughout the genome (cf. previous paragraph). It is sufficient that one of them is located in a microdeletion to indicate a loss of heterozygosity (hemizygosity) in the deleted region. In fact, in the absence of a microdeletion, the paternal and maternal

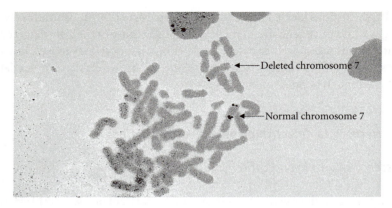

Fluorescence *in situ* hybridization of the metaphase of a child affected with Williams' syndrome with a probe specific for the deleted locus WSCR (Oncor). The deleted chromosome is recognized because of the telomeric control probe, the deletion for this syndrome is located just under the centromere of chromosome 7.

Figure 4.5   Microdeletion

alleles are found in a child; however, in the case of a microdeletion, the only markers detected will be those on the normal homologous chromosome (Figure 4.6). For the analysis of the microdeletion to be possible it is necessary that the markers used are sufficiently informative, and that the DNA of the affected child is studied at the same time as that of its parents. Above all, it is necessary to know and study the microsatellite markers in the deleted region and therefore suspect clinically that microdeletion.

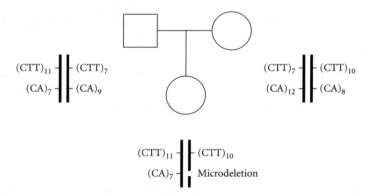

In this example, the use of a dinucleotide repeat polymorphism $(CA)_n$ allows detection of a *de novo* microdeletion of maternal origin; indeed only the paternal $(CA)_7$ is found in the child.

Figure 4.6   Molecular diagnosis of a microdeletion

4.4 Uniparental disomies

A person carries a uniparental disomy (UPD) when they have inherited the two homologues of a pair of chromosomes from a single parent, the other parent having not contributed anything to the pair considered (Engels, 1993, 1995). Obviously a standard karyotype analysis will show no anomaly, as all the chromosomes are normal. Uniparental disomy can only be detected when it is specifically looked for by a microsatellite analysis of the proband and their parents (Figure 4.7) which shows the absence of a contribution from one of the two parents for the pair of chromosomes considered. Several hypotheses can explain the origin of uniparental disomies; the classic is the return to disomy after a trisomic conception. In fact, it has long been known that aneuploid conceptions (in particular trisomies) are very frequent in humans, with the majority not being viable. They result from an error in mitosis at the level of the paternal or maternal gamete during the first or second meiotic division. After fertilization, the nonviable trisomies will be eliminated naturally unless a 'rescue' mechanism intervenes by the loss at random of one of the three homologous chromosomes, leading to a return to the disomic state. In between a third and a half of cases, the fetus will find itself with two homologous chromosomes originating from the same parent, and therefore carry a uniparental disomy. Sometimes a trace of the trisomy remains in the mosaicism of the placenta, which can be detected if a karyotype of the placenta is performed, and can be the source of complications of the pregnancy or retardation in intrauterine growth. If a non-disjunction, without crossing over, was produced in meiosis I, the pair of chromosomes will, one time out of three after rescue, be a uniparental pair (heterodisomy, Figure 4.8(a)). A heterodisomy can have clinical consequences that are more or less serious depending on the chromosome involved; in particular, they can be responsible for specific pathologies when genes

The use of dinucleotide repeat polymorphisms $(CA)_n$ and trinucleotide repeats $(CTT)_n$ allows detection of the case where the child has received two paternally derived chromosomes (heterodisomy) without any maternal contribution to the pair of chromosomes considered.

Figure 4.7 Molecular diagnosis of a uniparental disomy

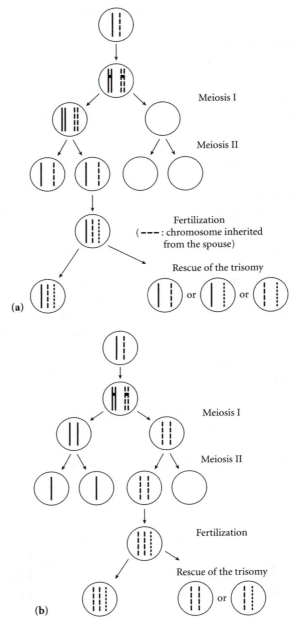

(a) *Heterodisomy*: distribution of a pair of chromosomes with non-disjunction at meiosis I. The 'rescue' of the trisomy by random loss of one of the three chromosomes can lead to the formation of a uniparental heterodisomy (in fact in a third of cases, the embryo receives two parental chromosomes without any contribution from the other parent of the pair of chromosomes).

(b) *Isodisomy*: distribution of a pair of chromosomes after non-disjunction in meiosis II. The 'rescue' of the trisomy by random loss of one of the chromosomes can lead in half of the cases to the formation of a uniparental isodisomy (the embryo has a pair of chromosomes derived from a single parental chromosome).

Figure 4.8 Heterodisomy or isodisomy: schematic representation of the generating mechanisms of a uniparental disomy, heterodisomy or isodisomy

which are under the control of parental imprinting are present on the chromosome. The best example is that of a uniparental disomy of chromosome 15 which is responsible for both Prader–Willi and Angelman syndrome according to whether the origin is respectively maternal or paternal. In contrast, if a non-disjunction without crossing over was produced in meiosis II, the pair of chromosomes will, one time out of two after rescue, be derived from a single parental chromosome (isodisomy, Figure 4.8(b)) and could possibly lead to the appearance of an autosomal recessive disease, the parent carrying the recessive allele transmitting a double dose of it by this mechanism.

It is important to notice that a non-disjunction in meiosis I or meiosis II, followed by a crossing over event can lead to a partial heterodisomy for the genes telomeric to the crossover, and a partial isodisomy for the genes centromeric to the crossover (see Figures 4.8(a) and 4.1(d) for the first case and Figures 4.8(b) and 4.1(d) for the other).

Finally, a uniparental disomy of certain chromosomes can also lead to no particular pathology. The most difficult problem is to recognize the indications leading to a specific search for a uniparental disomy and on which chromosome, because there is no method available to search them all systematically. Disomies are probably quite rare but they should be suspected starting from the cytogenetic data (existence of mosaicism or a Robertsonian translocation for example), family history or from specific echographic signs (Engels, 1995).

4.5   Conclusions and perspectives

The standard karyotype analysis remains the golden standard for visualizing the collection of all chromosomes and enabling the detection of a large majority of chromosome anomalies. However, molecular biology is becoming more and more like an indispensable complement bringing a useful rapidity to urgent tests of anomalies of number, and above all anomalies undetectable from the karyotype. The description of clinical syndromes associated with microdeletions, duplications and uniparental disomies will, in the years to come, become more precise and wider in scope, and the methodologies for diagnosing them will become more and more indispensable. The development of DNA chips should, for example, allow the use of a method for testing a very large number of chromosome microdeletions and duplications located in the telomeric regions at the same time.

4.6   References

Antonarakis, S. E. (1991). 'Parental origin of the extra chromosome in trisomy 21 as indicated by analysis of DNA polymorphisms.' The Down Syndrome Collaborative Study Group. *N. Engl. J. Med.*, **324**, 872–876.

Boué, A. (1989). *Médecine Prénatale*. Flammarion Médecine-Sciences.

De Grouchy, J. and Turleau, C. (1982). *Atlas des Maladies Chromosomiques*, 2nd edn. Expansion Scientifique Française.

Engels, E. (1993). 'A new genetic concept: uniparental disomy and its potential effect, isodis-omy.' *Am. J. Med. Genet.*, **46,** 670–674.

Engels, E. (1995). 'La disomie uniparentale: revue des causes et conséquences en clinique humaine.' *Ann. Genet.*, **38,** 113–136.

Gardner, R. J. M. and Sutherland, G. R. (1996). *Chromosome Abnormalities and Genetic Counselling*, 2nd edn. Oxford University Press.

Lejeune, J. *et al.* (1963). 'Trois cas de délétion partielle du bras court d'un chromosome 5.' *CR Acad. Sci.*, **257,** 3098–3102.

Muller, F. *et al.* (2000). 'Parental origin of the extra chromosome in prenatally diagnosed fetal trisomy 21.' *Hum. Genet.*, **106,** 340–344.

5 Screening and identification of pathogenic and exogenic agents

Nevine Boutros, *Hôpital Saint-Vincent de Paul, Paris*
Anne Casetta, *Hôtel-Dieu, Paris*
Denis Cointe, *Hôpital Antoine-Béclère, Clamart*
Annick Diolez, *Institut national de la recherché agronomique
Versailles*
Liliane Keros, *CHU Paris-Sud*

5.1   Clinical virology

5.1.1   Introduction

Viral diseases have been known for thousands of years since the clinical characteristics of smallpox were described by the Chinese around 2500 BC. In the 18th century, Jenner discovered that infection by vaccine protected humans against smallpox. At the end of the 19th century, the virus was defined as an infectious agent responsible for transmissible diseases, characterized by their invisibility under an optical microscope, and their filterability using a porcelain membrane. In the 20th century considerable progress in the knowledge about viruses and viral infections was made leading to the possibility over the last few decades of diagnosing these infections or, in a broader sense, the performing of medical virology analyses. These analyses have for their goal:

- to prove that a virus is responsible for an acute infection;

- to associate a virus with a chronic infection;

Diagnostic techniques in genetics   Edited by Jean-Louis Serre
© 2006 John Wiley & Sons, Ltd

- to determine the immune status of an individual, by checking whether the individual has been in contact with a virus without considering the date of infection;

- to follow the evolution of a viral infection;

- to study the epidemiology of viral infections.

5.1.2   Classical methods of analysing viral infections

Until the end of the 1980s, only the so-called classical methods were used in the medical analysis domain. Despite the development of molecular biology, these techniques are still very much in use today; this is why, in this chapter, the classical techniques will be described first, followed by the advantages and disadvantages of both techniques.

Classical direct methods for virus detection

These methods are based on the detection of viral antigens. Viruses are obligatory parasites that can only multiply in living cells. Today, only cells in culture are used in normal practice from embryonic chicken eggs or laboratory animals relevant to the search.

In cell culture, characterization of a virus can be done by observation of morphological modifications of the cells (cytopathic effect) or by the characterization of viral antigens. Viral antigens are recognized by specific antibodies; these can be polyclonal antibodies, usually obtained by experimental immunization of animals or, more frequently nowadays, monoclonal antibodies, most frequently from mice. The interest of the latter rests in the precision of the recognized antibody structure.

The visualization of the antigen–antibody complexes can be performed using various procedures: agglutination of sensitive particles, detection of fluorescence or coloration due to the antibody being marked by a fluorochrome or an enzyme respectively.

Indirect methods

These methods are based on the detection of antibodies synthesized in response to a viral infection. According to the goal of the research, it is possible to consider the total antibody population or a particular class of antibody (IgG, IgM, less frequently, IgA). The IgM antibodies appear first, on average about 2 weeks after the contagion (contact with a contaminated subject); they disappear at around 2 months. The IgG antibodies appear several days after the IgM antibodies and last, in general, for life. In the case of infections acquired *in utero*, the IgM antibodies can be detected in blood from the umbilical cord from the 22nd week of amenorrhea. In practice, it is not possible to distinguish between IgG antibodies transmitted passively from the mother to the fetus from IgG antibodies synthesized by the fetus.

The main techniques to detect antibodies in use at the moment are the immunoenzymatic techniques (ELISA), which are based on the detection of antigen–antibody complexes with the help of enzyme-marked antibodies.

Uses of classical methods

The classical methods have a number of advantages in common and some particular advantages depending on the method used. As a general rule, the classical methods are simple to set up and are relatively cheap.

The infection of cells in culture is the only technique that permits isolation of a virus and the study of its sensitivity to antiviral agents. Detection methods for viruses that are based on the detection of viral antigens have the advantage of a rapid response in comparison with culture-based approaches. Indeed, several days or weeks are necessary to detect a virus in culture, while detection of viral antigens can be performed in several hours.

Only serological methods, which have as their goal the detection of antibodies, permit establishment of the immune status of an individual. This property is important to consider because, for infections such as rubella or hepatitis B, it can be discovered if an individual is protected against the infection (presence of antibodies), or if it is necessary to vaccinate (lack of antibodies). In addition, determination of the class of antibodies allows differentiation between a primary infection (first contact with the virus) and a secondary infection (reinfection, replication). This is particularly important in certain circumstances, notably in pregnant women. In fact, infections which can be transmitted from the mother to the child *in utero* (cytomegalovirus, parvovirus B19, rubella and chickenpox) have much more serious consequences for the fetus in the case of a primary infection than with a secondary infection.

Disadvantages of classical direct methods

Methods for virus detection by cell culture or by the detection of viral antigens have the major disadvantage of low sensitivity. This is a real problem, notably in patients who have received grafts for whom it is very important to identify an infection early so that it can be treated as rapidly as possible. In the same way, it is very difficult to make a diagnosis of herpetic meningitis by looking for the virus in the cerebro-spinal fluid by culture methods, because of the low concentration of the virus in the biological medium. These methods allow diagnosis of an active infection, but do not allow a primary infection to be distinguished from a secondary infection.

Culture-based methods have some specific disadvantages.

- Firstly, and very importantly, several viruses cannot be cultivated (certain enteroviruses), or are difficult to cultivate (parvovirus B19, the rubella virus).

- The viruses must be 'alive', but viruses possessing an envelope are fragile and degrade very rapidly. In these conditions, it is necessary that the samples that potentially contain such viruses are stored at a very low temperature ($-80\,^{\circ}$C) in order to obtain reliable results.

- Despite the presence of antibiotics in the culture medium, the possibility of bacterial contamination is always present.

- The length of time required for the detection of certain viruses by culture-based methods can lead to a delay in the diagnosis of a viral infection.

- The difficulty of culturing certain viruses, as well as the low sensitivity of this method, means that apart from the diagnosis of cytomegalovirus infection, culture-based methods are not well adapted to antenatal diagnosis of fetal infections.

- Finally, culture-based methods give little quantitative information about viral levels.

Disadvantages of classical indirect methods

On average, antibodies cannot be detected until about 2 weeks after the start of the infection, which greatly delays diagnosis.

As was seen before, it can be important to distinguish between a primary and a secondary infection. The diagnosis of a primary infection rests on the observation of a seroconversion (appearance of antibodies), or the detection of both IgG and IgM antibodies. Seroconversion is, in fact, rarely observed as it is only detected in the case of a systematic screen for infection. At the moment when the clinical signs are present, the antibodies are generally already detectable; in addition, certain infections are asymptomatic and can pass completely unnoticed. The presence of IgM antibodies is often difficult to interpret without an evocative clinical context. IgM antibodies are, in fact, always detected during a recent primary infection if sufficiently sensitive methods are used, but they can also be detected a long time after the start of a primary infection (persistence of IgM), during a secondary infection, because of a cross reaction between the pathogenic agent and another virus, or because of a polyclonal non-specific stimulation of the immune system due to other infectious agents. The latter reason is very likely to be the most frequent cause of the detection of specific IgM antibodies not linked to the studied disease.

It is theoretically possible to detect IgM antibodies in fetal blood from the 22nd week of amenorrhea. This presence signifies a fetal infection because maternal IgM antibodies cannot be transmitted to the fetus. In fact, only infection with rubella can be reliably detected by this method. For other infections, the detection of IgM antibodies is random. In the same way, after the birth it is not possible, apart from for rubella, to use this method to make the diagnosis of a congenital infection.

Finally, serological methods do not allow the evaluation of the seriousness of a viral infection in any situation.

5.1.3   Analysis methods for viral infections using molecular biology

Reasons and limits of success

Molecular biology techniques have quickly spread into virology practice. They have radically modified the care for diseases such as that due to human immuno-defiency virus (HIV) or the hepatitis C virus (HCV). Because they use samples as inert media (not requiring culture), molecular biological techniques are less subject to the major constraints of the classical analysis methods (see above). This is the reason for the rapid spread of molecular biology methods in virology practice. In addition, the extremely detailed genetic information, previously unavailable, on the infecting viruses given by these methods assures that their use will become more and more widespread.

However, apart from the general limitations which slow down the expansion of the use of molecular biology techniques (cost, size and weight of the equipment to be put in place, extreme sensitivity that opens the door to contamination), these techniques have two significant disadvantages for their use in virology.

Genetic variability   Viruses are capable of extreme evolutionary adaptation that translates into a high level of plasticity of the genetic material. The mutations, more or less frequent depending on the virus and the genomic regions, appear during replicative cycles. However, all of the molecular biology techniques depend for at least one of their steps on hybridization, determined by the complementarity between the viral sequence and an oligonucleotide. A large amount of genetic variation imposes strong constraints on the choice of sequence and size of the oligonucleotides. These limitations are reinforced when an amplification step is performed because a single mismatch at the 3′ end of the primer can prevent the initiation of the reaction. The consequence of the high viral variability is that the choice of techniques and primers has to be made with great care, using data from sequence banks. The result is that certain techniques, like ligase chain reaction (LCR, see Chapter 1) or differential amplification of certain viral sub-populations by selective priming, are difficult to set up for a given virus, even in a region of supposedly low variability. In certain circumstances, it is wise to perform multiple amplifications to detect a virus with extreme variability. The effects of the variability are not limited just to the hybridization region. In general, it can be considered that techniques based only on hybridization are more resistant to the constraints imposed by viral variability, since the oligonucleotides will have been selected with some care.

The sensitivity threshold   While the most sensitive amplification-based methods are capable of detecting between one and five copies of target sequence per reaction, the effects of earlier steps (volume of the test sample, extractions, possible reverse transcription) means that the minimum sensitivity threshold is actually between 20 and 50 genome equivalents per millilitre (ml) of sample or for 10^5 cells. The usual thresholds for amplification methods are 10 to 50 times higher and methods based only on hybridization, except for the DNA branching method, are more than 1000 times less sensitive. The temptation to conclude that it is preferable to use the most sensitive method is a mistake. For viruses (such as the rubella virus) that are present in certain biological liquids in concentrations lower than 10^2 to 10^4 genome equivalents/ml, a highly-sensitive method is justified. Conversely, for viruses commonly found at 10^6 or 10^8 genome equivalents/ml like the hepatitis B virus (HBV), it is prudent to use a more robust method, less sensitive to contamination. For HBV, the choice of technique is radically different when tracking a residual replication during or after specific antiviral treatment or when it is a question of dealing with the safety of transfusions, than when (for technical reasons) it is necessary to perform the analyses on serum pools, requiring the technique with the maximum sensitivity. In the same way, the choice of a detection method specific for cytomegalovirus (CMV) is made using different criteria if it is a question of detecting a viral replication in an immuno-suppressed patient or to confirm the existence of a fetal infection. The sensitivity threshold of the molecular biology methods used should be adapted to the clinical question.

Methods to detect a viral presence applied to acute diseases

In this context, molecular biology methods are used whereas the indirect diagnostic methods are no longer usable. Often, the detection of a viral marker during an acute phase of an infection, when there are characteristic symptoms, is enough to confirm the viral origin of the disease. Theoretically possible for all viruses, diagnostic approaches using molecular biology techniques are actually used for potentially serious diseases or in the following circumstances.

(a) When reagents for a recently described virus allowing confirmation of viral in- fection by a less onerous method (monoclonal antibodies for the direct methods, antigens for serological analyses) are not yet widely available, as for the search for infection by the virus responsible for hepatitis E. It is then the flexibility of the setting up of molecular biology methods that makes them successful, because they 'only' require knowledge of the genomic sequences and their variability, and to have proven experience.

(b) When an easy to use direct diagnostic method does not exist for some well known viruses, while the clinical symptoms are very suggestive, so that a positive result allows the interruption of a difficult etiological inquiry to detect an underlying genetic pathology, and would justify the use of a treatment which would make the direct approach ineffective (i.e. acute non-regenerating anaemia due to parvovirus B19). It is because they are easily applicable when serological analyses are not yet available or are not informative that these techniques are developed.

(c) When the direct diagnostic methods are ill-suited or are judged unsatisfactory, while the disease is particularly severe or acute and requires quickly putting in place specific therapeutic and preventative measures. Examples would be the rabies virus or, for a patient affected with fulminating hepatitis, the search for hepatitis δ. In both cases, the classical reference techniques can only give results too late to be medically useful. It is the speed with which molecular biology techniques can give a result that counts. Occasionally it is also the high sensitivity of the molecular biology methods that justifies them being put in place. In this category, examples are herpetic meningoencephalitis (for which viral culture is not sensitive enough and the detection of specific immunoglobulins in the cerebro-spinal fluid too slow) or the search for respiratory syncytial virus in the case of bronchiolitis in infants (for which there exists rapid and reliable direct methods).

(d) When the clinical symptoms that lead to the viral inquiry are caused by a limited number of viruses (use of group or 'multiplex' amplifications).

In conclusion, the techniques of molecular biology used during acute infections enter into the collection of direct diagnostic methods. Their flexibility in setting up, their great sensitivity (which can also lead to the use of more accessible samples) and their capacity to give results quickly are the reasons for their success. As techniques for direct diagnosis, they only give results in a window of time limited to the productive phase of the infection, whether a primary infection, a reinfection or a replication. Because they are more sensitive than classical methods, the duration of detection of viral replication is increased. However, because of the diversity of methods used, the sensitivity thresholds are variable and the windows of use of these methods cannot always be determined exactly.

Methods to detect a viral presence applied to chronic diseases

Aside from the case of a major immune deficit, the detection of specific antibodies indicates the existence of a previous contact with the virus and, in general, gives little information about the current state of the patient. Serological analysis is of little use in the case of very common infections (such as infection by the Epstein–Barr virus or EBV) or where it is difficult to distinguish the medically interesting subgroups (such as for the papillomaviruses, which are potentially oncogenic during viral infection of the cervix). In rare cases (HBV), serological analysis can establish the existence of an incomplete immune response showing a chronic or persistent infection. The direct diagnostic methods fit downstream of the serological approach and give more information about the link between the virus and the pathology. Therefore, in the case of a hepatitis B infection, having established by serological techniques that the patient is infected, estimation is then made of whether persistent viral multiplication exists by a qualitative research for the virus in the plasma.

As a general rule, detecting a virus associated with a chronic infection in a qualitative manner is not enough to establish its responsibility for the infection. A critical analysis of the clinical, biological and epidemiological data is required. The interpretation of the results and the selection of the molecular biology methods to be used in the search for a virus responsible for a chronic infection will be influenced by the host, the virus, the pathology considered, the nature of the sampling and the technique. This step will probably succeed in identifying the virus responsible for the clinical symptoms. Sometimes, it is useful to measure in a subsequent step certain quantitative characteristics like the 'viral charge' for infection with the hepatitis C virus (HCV) and/or qualitative characteristics like the type of papillomavirus in cervical dysplasias or the genotype of the HCV virus which can affect the choice of therapy.

In practice, the high sensitivity of molecular biology techniques is used in chronic persistent infections:

- in the case of a negative result, to verify the eradication of the viral infection by the immune system or by therapy;

- in the case of a positive result, to justify a complementary inquiry and a clinical follow-up that will give more information about this chronic infection.

Particular applications in clinical virology

The molecular biology methods used can differ depending on whether a DNA or an RNA virus is being studied. The methods of sample amplification can be distinguished from the methods of signal amplification.

Sample amplification methods   Methods based on iterative amplification of
DNA molecules (PCR, LCR) can be distinguished from those based on iterative
amplification of RNA (NASBA, see Chapter 1). The former allow the detection of
the genome of a DNA virus and require adaptation through the addition of a preliminary reverse transcription stage to be able to detect RNA viruses. The latter, which
in theory are not constrained by the nature of the initial genetic material, are most
often used on RNA samples. They are used in virology for the quantification of RNA
viruses or for the detection of transcripts associated with the replicative cycle of a
DNA virus, either latent or reactivated.

Signal amplification methods   Essentially based on a single hybridization
(branched DNA method, Qβ replicase, RNA/DNA hybrid detection), the signal amplification methods are effective for the detection of viral genomes whatever their
nature.

Examples   The following three examples show the special use of these methods in
virology.

Herpetic meningoencephalitis. A primary herpetic infection or a recurrence of herpes in a fever context can produce a set of symptoms which can lead to the suspicion
of viral meningoencephalitis. To establish the diagnosis quickly justifies the maintenance of an intravenous antiviral treatment for 2 to 3 weeks, based only on the clinical
criteria and, in the case of a negative result, to continue the diagnostic investigation.
   The other viruses in the herpes group can be associated with encephalitis, meningoencephalitis, with medullary attacks or with cerebral lymphomas (VZV, CMV,
EBV, HHV-6, HHV-7). Depending on the clinical cases with which they are faced,
the laboratories can be pushed to widen their diagnostic strategy. They could keep a
PCR strategy capable of detecting all viruses in the herpes group, and therefore avoid
performing a specific analysis for each virus. This is done either by targeting a conserved region amplifiable by a single pair of primers (group PCR), the identification
of the virus responsible obtained by secondary characterization of the amplicon, or
by performing a series of PCR reactions in a single tube, each specific for a given
virus so the amplification products can be simply distinguished by electrophoretic
migration (multiplex PCR).

Detection of enteroviruses. The enteroviruses form a heterogenous sub-group of
the picornaviridae family. The name covers diverse groups such as polioviruses
(three types), group A coxsackieviruses (23 serotypes), echoviruses (31 serotypes),
enteroviruses 68 to 71 and hepatitis A virus (enterovirus 72). The enteroviruses,
grouped according to a common genetic structure with a marked variability, show a
wide divergence in their pathogenic capabilities. In general, they are responsible for
epidemics of classically resolved acute meningitis, without side effects, when caused
by echoviruses, coxsakievirus B and, rarely, coxsackievirus A. Their detection in the

cerebro-spinal fluid is problematic due to the wide range of viruses looked for, with a given virus having high levels of genetic variability. The target 'consensus' regions of the primers that can be used for detection by single or nested RT–PCR are highly constrained and even so can contain degenerate sequences. An original strategy for making primers has been developed using 'stair primers'. With this method, in a given low variability region, the effects of the residual variability can be reduced using a mixture of overlapping primers where the 3' end is shifted progressively one base at a time. This strategy reduces the effects of viral variability, however, depending on the virus and the primer mixture, it can have variable detection thresholds and it can be difficult to reproduce an identical mixture.

Selection of amplification controls. In the case of negative results, it is necessary to consider the entire process, the type of test and sampling procedure, going back to the study of the symptoms, whether the methods used for the clinical analysis were sensitive enough, and whether the analysis was performed correctly. The amplification controls commonly used in PCR analyses to test the success of the extraction and amplification steps and the absence of common causes of PCR problems (single copy genes, mRNA of 'house-keeping' genes) are often non-operational in virology, with analyses performed on acellular fluids or windows of detection of the controls disproportionate with respect to the number of viral particles that could be present in the sample. It is necessary to use a marker quantity that is close to the detection threshold of the method, so that it has the same detection constraints as the virus. The best situation would be to have a control almost identical to the virus being investigated, added to the sample, which would then pass though all stages of the analysis procedure.

Quantitative methods allowing the control of the infection

These methods are set up, at least in theory, after the demonstration of the presence of the virus using direct methods. The reasons and modalities for their use vary for each virus. For practical or historical reasons, the detection and characterization of the virus can be combined in a single step, as in the case of certain quantitative techniques. In general, the aim is to establish the parameters allowing guidance of the therapy, to evaluate the prognosis for the infection, and to give more information about the epidemiology and chains of transmission. It is necessary to separate quantitative methods from qualitative methods.

Interpretation principles and criteria   The quantity of virus present in a tissue, an organ and an organism is the result of its capacity for replication and the immune and/or therapeutic response that limits replication. For the moment, with respect to acute infections, use of virus quantification correlated to some symptoms is neither practical in general, nor usual. It is only possible to consider in the long term the quantity of virus as a factor in the severity of an infection allowing a statistical prediction of associated after-effects, in the case of herpetic meningoencephalitis, or during a fetal infection by CMV.

For viruses responsible for chronic latent diseases that are susceptible to reactivation, like CMV and EBV, quantification by molecular biology methods can provide thresholds above which the presence of a virus is clearly associated with symptoms (current or future), thresholds below which the hypothesis that a virus is the causal agent of a pathology must be rejected, and a vast grey area where interpretation is more complex.

For viruses responsible for chronic persistent diseases, quantification does not allow at a given moment the establishment of a strict link with a precise set of clinical symptoms. It can allow evaluation of the efficacy of an on-going treatment with a viral replication inhibitor for HIV and HBV, or the chances of success of a planned treatment for HCV. In a prospective manner, it can serve, among other factors, as a predictor of the evolution towards a more serious form of the disease (HIV) or of the transmission risk (HIV, HBV, HCV), therefore justifying the introduction of preventative strategies.

Application of viral quantification   Quantification is performed by methods that must be adapted to the viral titres spontaneously present in biological fluids. It can be performed using hybridization, detecting hybrids in the normal way using radio-isotopes (I^{125} for HBV), or through using a signal amplification technique (DNA–RNA hybrid method for HBV and CMV, branched DNA method (bDNA, see Chapter 1) for HBV, CMV, HIV and HCV).

Quantification can also be performed using substrate amplification techniques. Those based on the iterative amplification of DNA or RNA molecules, semi-quantitative methods where a signal obtained after amplification is compared with that obtained using an external range standard, are no longer used. They have been replaced by competition-based methods where a wild-type genome and a standard are amplified in the same test tube. This can be done by evaluation at the end of amplification, or during the exponential phase ('real-time' quantification, see Chapter 1); the latter is for the moment reserved for PCR-based methods.

Qualitative methods allowing the control of the infection

Interpretation principles and criteria   Qualitative analysis of an infection is, as a general rule, performed in the case of chronic diseases. It can be performed in order to account for the failure of a previously completed course of therapy (the search for viral resistance), or to predict the chances of success of a future course of therapy (the search for mutations associated to the resistance to antiviral agents, HCV genotyping). Some qualitative information can also be indispensable, when it is a question of analysing the modes of transmission of a virus from one patient to another, or of a virus in a hospital ward or other location. These epidemiological approaches are often in the domain of medical–legal practice or scientific research. From the point of view of interpretation of qualitative methods, it is necessary to distinguish the methods used for a prospective study from those that describe (a) a viral population present at the moment of the analysis and arrive at a direct conclusion such as the search for resistance to an anti-viral agent, or (b) the similarity (estimated using criteria that are sufficiently discriminating) of two strains isolated from patients implicated in the same chain of transmission. The former put diagnostic virology in the established domain of predictive viral genetics. Their development has been/is founded on statistical data or even on evolutionary physiopathological models.

Qualitative methods in practice   The acquisition of qualitative information on the infecting virus can be performed in the following ways.

- *By differential amplification:* with the help of specific primers capable of distinguishing different viral sub-types (initial methods of typing HCV and HCB, certain methods of sub-typing HIV).

- *By enzymatic digestion followed by electrophoretic separation:* (a) the amplicon segments resulting from differential truncation performed with the help of restriction enzymes that recognize some characteristic sequences (RFLP analysis), (b) segments derived from whole genome RFLP analysis of the strain (Figure 5.1).

- *By differential hybridization of the amplification products:* these methods are performed with the help of selective probes targeting the characteristic sequences followed by detection of the results of hybridization. Techniques based on probes of restricted size are sometimes not well-adapted for viruses with marked variability (HIV).

- *By differential migration:* in native gel, of single-stranded DNA obtained, after dilution and denaturation of the amplicon, followed by guided reannealing so that the separated strands fold up on themselves according to the constraints imposed by their sequence (SSCP technique). These methods are mostly used when, for epidemiological reasons, it is desired to distinguish many isolates that are very similar with marked genetic stability.

- *By direct sequencing of the amplification product:* these laborious methods provide a lot of information, but their use requires the capability of finding those data of medical interest within this large amount of information.

- *By construction of recombinants:* these methods can be used where a database – sufficiently complete to allow the reliable interpretation of sequencing results – does not yet exist, and isolation of the original strain is difficult or impossible.

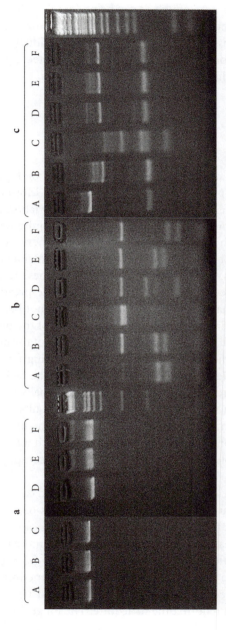

The amplicons from amplification by RT-nested PCR of the genome of the viruses contained in the serum of six patients (A–F) are not digested (a) or well digested by the recommended mixture of *HaeIII-RsaI* restriction enzymes (b) and *Bst* NI-*Hinfl* (c). The restriction fragments are separated by electrophoretic migration. The comparison of the digestion profiles thus obtained with the expected theoretical profiles allows determinations of the genotypes of the infecting strains (six genotypes numbered from 1 to 6 have been indexed for HCV).

Results in order: patient A, genotype 2 - patient B, genotype 4 - patient C, genotype 1 - patient D, genotype 3 - patient E, genotype 4 - patient F, genotype 3.

If necessary and according to the genotype established by the first analysis, additional digests are performed following the same principle, with the help of other selected restriction enzymes, permitting sub-typing of the strain. The intratypic diversity can give distinct restriction profiles for the same HCV genotype (e.g. genotype 3 of patients D and F).

Figure 5.1   Genotyping of the hepatitis C virus (HCV) by RFLP

Another way of characterizing an infection uses markers of a particular replication cycle in a given tissue type. Thus, for certain DNA viruses responsible for chronic reactivatable latent infections, with certain samples such as blood, demonstrating the viral genome in leucocytes is not enough. It can be preferable to look for viral mRNA transcripts specific for an active infection using RT–PCR or NASBA. In a similar fashion for HIV, it is possible to look for markers of residual replication in peripheral circulating lymphocytes, in the ganglions, while detecting and even quantifying there the mRNAs that code for viral transactivators, by quantifying the genomic RNA.

5.1.4   Conclusions

In this chapter, several of the details of the practice of molecular biology applied to diagnostic virology have been presented in a succinct manner. Molecular biology methods have extended the diagnostic possibilities of virology laboratories, by substituting for laborious procedures poorly adapted to clinical requirements, simplified strategies more in tune with the requirements of the area. They have enabled taking control of diseases caused by certain viruses that are inaccessible using the other direct diagnostic techniques. It is, among other things, for these reasons that in virology the techniques have been adopted in an early and widespread fashion. It is, however, necessary to note that their virtue of simplicity had not been pushed as far as in other diagnostic domains, in part due to the diversity of the pathogenic agents, the variability of the clinical situations, and the multiplicity of the host–virus relationships which exist. Even when it is 'just' a question of testing for the presence of the virus, a certain complexity remains, not so much in the technological requirements too often put forward as in the choice, in an occasionally urgent context, of a well-adapted procedure and in the interpretation of the results, notably for chronic infections. From this point of view, it is not certain that the uniformity of analyses associated with the availability of diagnostic kits (for which viruses, which diseases, validated for which samples?) is sufficient to resolve these complexities and diversities.

Because in virology the practice of molecular biology is 'old' and can sometimes give insufficient results and because the knowledge and practices of virology are extending, approaches which establish viral parameters and that can be used to guide the therapeutic course of treatment are being used more and more frequently. They use a very varied set of methods, even if some are going to fall into disuse. Above all, they do not follow the simplifying tendency. They sometimes rely on a high level of specialized knowledge, use complex strategies and need for their interpretation large databases that require time to acquire. The longevity of each of these methods depends on its informativity and on its degree of adaptation to the practical constraints imposed by operational virology. It also depends on the capacity of virologists to use difficult methods that possess a certain degree of inertia with respect to adaptation, for the study of subjects where certain aspects can evolve very rapidly.

5.2 Clinical bacteriology

5.2.1 Introduction

In the domain of bacteriology, the classical techniques of detection and identification of microorganisms in clinical samples remain the 'gold standard' by virtue of their ease of use and interpretation as well as their low cost. The technological revolution of the last few years, with the appearance of molecular biology techniques, has nonetheless improved the classical approach, when it has been shown to be necessary in domains as varied as taxonomy, bacteriological diagnosis, detection of antibiotic resistance and bacterial typing.

5.2.2 Bacterial taxonomy

Taxonomy is the science of classifying living organisms. It is the practice of putting organisms into groups. The species is a concept that appeared, for bacteria, with the improvement in the knowledge of physiology from molecular biology. In microbiology, the aim of classification is to summarize all of the properties of a group of bacteria. The assigning of a bacterium to a given group enables the prediction of its properties without having to completely explore them. This predictive aspect is very useful to clinicians, who are passive users of bacterial classification, because it permits them to deduce all of the characteristics of a bacterium once it has been assigned to a group.

The identification of a bacterium is traditionally done starting with a certain number of tests concerning morphology, characteristics of their biochemistry and culture. These traditional 'hierarchical' classifications, characterized by a chosen order of the characters, are currently being called into question. The development of informatics methods has allowed the appearance of numerical taxonomy that helps take account of all observed characteristics without giving them a particular value individually. Numerical identification uses visual or automatic reading of biochemical tests.

The current revolution in taxonomy is based on the development of phenetic and genetic methods. The phenetic and, above all, the genetic criteria allow the precise definition of a bacterial species. A study restricted to a single phenotype has the drawback that a single error in a main pathway can cause the entire identification to collapse. Molecular and genomic analyses of bacteria concern the physio-chemical nature of their genomes, and look notably at the estimation of the percentage of [guanine+cytosine] nucleotides in the genomes, their degrees of homology, their size and their diversity, but these analyses can only be performed in laboratories which specialize in bacterial taxonomy.

The [G+C] percentage is the number of base couples guanine + cytosine per 100 base couples in a bacterial DNA molecule; it is measured by high-performance liquid chromatography (DHPLC, see Chapter 1). The G/C percentage varies between 25 per cent and 70 per cent in the majority of studied bacterial species. For example, its value is 32–36 per cent in *Staphylococcus aureusi* and 50 per cent in *Escherichia coli* K12.

The degree of homology between two DNA bacteria is evaluated by DNA/DNA hybridization from the fraction of genomes susceptible to form duplexes in given conditions of ionic force and temperature. It is considered that two bacteria are from the same species if their genomes hybridize at more than 70 per cent.

5.2.3   Bacteriological diagnosis

Clinical bacteriology is the study of the bacteria responsible for infections in humans. The nature of these infections can be diverse: respiratory, urinary, genital, etc. The search for the agent responsible for the infection requires a procedure with several steps, from the isolation of the bacterium in the pathological sample to the identification of the bacterial species. Initially, a direct examination of the bacterial sample is performed using Gram staining to differentiate Gram positive bacteria (coloured violet), from Gram negative bacteria (coloured pink), or Ziehl–Neelson staining for mycobacteria, which are acid–alcohol resistant. After this first step, the bacteria are cultivated on solid or liquid gel medium. The delay for culture is usually 24 hours, but can be between several days and several weeks depending on the species, after which a complete identification is undertaken. While bacteriological diagnosis is still currently based on 'classical' techniques, molecular biology methods allow in certain cases more rapid or more precise identification. Direct quantitative or qualitative testing for bacterial nucleic acids (DNA and RNA) in the clinical samples can be performed. However, for rapidly growing bacteria, this detection mode is limited to a few specialized laboratories. Its use is restricted to a few very particular cases such as the testing using PCR of *Neisseria meningitidis* in the cephalo-spinal fluid when the culture is negative in a patient on antibiotics. In contrast, for slow growing bacteria, or for those requiring sophisticated culture techniques (cell cultures), molecular biology techniques have been developed in several laboratories. Some examples are described below.

The mycobacteria

Mycobacterium tuberculosis is very important on the global scale with around 10 million new cases each year. Pulmonary tuberculosis is the most frequent clinical form and kills more people in the world than all of the infectious pathologies due to the other microorganisms.

For many years, the only rapid diagnostic test for bacterial tuberculosis was the microscopic examination of sputum using Ziehl–Neelsen staining. The specificity (capacity to specifically identify a given bacteria) of this technique is good, close to 95 per cent, but the sensitivity is low (40–60 per cent) as bacteria are emitted in an intermittent fashion that makes repetition of the sampling indispensable. With either a positive or negative microscopic test, the definitive diagnosis rests on the isolation in culture of mycobacteria from the biological samples (sputum, gastric tubes, urine, etc.) on special media. The sensitivity and specificity of cultures is higher. The principal problem comes from the delay for a positive test, which varies from 3 to 50 days, depending on the number of bacteria found during the direct examination.

Since their introduction, molecular identification techniques have had a large amount of success and their application to this problem has rapidly become widespread because they normally allow results to be obtained in several hours.

Mycobacterial DNA is directly looked for in respiratory samples using PCR. Several kits are commercially available in Europe. The sensitivity of this method is strongly influenced by the concentration of bacteria in the sample. A sample with few bacteria can lead to a false negative result. On the other hand, there can be the fear of a false positive result resulting from possible contamination of the sample by exogenic DNA, produced by previous or concurrent reactions. In terms of sensitivity, the detection of *M. tuberculosis* in a sample remains less than that of culture. For these reasons, the place for molecular biology in bacteriological diagnosis is still a controversial subject.

However, molecular biological techniques have been developed with the goal of differentiating between different clinically interesting species of mycobacteria. Currently, certain commercially available kits allow distinction and identification of *M. tuberculosis, M. avium, M. intracellularae, M.kansasii* and *M. gordonae.* This test is based on the use of DNA probes marked with acridinium ester (luminescent molecule) that specifically complement the different types of bacterial ribosomal RNAs. The results are available in around 1 hour. The interest in these identification probes rests in the interpretation of positive cultures from clinical samples grown on liquid or solid media in immuno-suppressed patients where a pulmonary infection may be due to so-called 'atypical' mycobacteria (*M. avium/intracellulare, M. kansasii,* etc.).

Chlamydiae

Chlamydiae are bacteria that have an obligatory intracellular development with a particular cell wall structure (absence of peptidoglycan). Four of the known species are pathogenic in humans, of which *C. trachomatis* and *C. pneumoniae* are responsible for urogenital and respiratory infections respectively.

C. trachomatis is the main worldwide bacterial cause of sexually transmitted diseases (STD) and has a high morbidity if it reaches the upper genital tracts of women.

It can also be a cause of tubal sterility and increases the risk of extra-uterine pregnancies. The biological diagnosis of this organism is difficult because *C. trachomatis* cannot be cultivated in normal media. The diagnostic method used up to the present (direct immuno-enzymatic and serologic test) is not very sensitive, or specific (cross-reactions between different species) and are difficult to perform. The reference test is still cell culture, which demands fastidious care of the cells as well as a delay of 3 to 7 days before obtaining a result. Some molecular biology techniques have been developed to reduce the delay necessary for bacteriological diagnosis and facilitate the test procedure. The most used technique is PCR which has the advantage of having a high sensitivity (stability of nucleic acids in the sample and detection of a small number of bacteria) and a high specificity (choice of probes, reduction of false positive reactions by inclusion of an anti-contamination system), as well as results which are reliable and reproducible. These gene amplification techniques are more sensitive than isolation in cell culture (the reference method), which has a sensitivity of 80 per cent.

    C. pneumoniae is responsible for a diverse range of respiratory infections: pneumonia, bronchitis, sinusitis and pharyngitis. The direct or indirect detection of *C. pneumoniae* is difficult because cell culture has low sensitivity, is long, difficult and rarely positive. The direct search for *C. pneumoniae* in clinical samples by using fluorescently labelled antibodies has a subjective interpretation. In addition, interpretation of serology is made difficult by the absence of seroconversion (increase in antibodies in the body of the patient), the presence of cross-reactions with other species and raised seroprevalence (certain people keep the antibodies for a long time). As for *C. trachomati*, the diagnosis can be performed using gene amplification. The nucleic acids are detected in pharyngeal samples. The detection of DNA from *C. pneumoniae* in these samples without an accompanying serological response (appearance or significant increase in antibodies) could be due to a prolonged carrying period rather than an acute infection. The development of a quantitative PCR technique could allow better distinguishing of carrier and infection status (see clinical virology).

Mycoplasmas

Mycoplasmas are mollicute bacteria, which are distinguished from other classes of bacteria by the absence of a cell wall. This organism cannot be seen on coloured smears and cell culture requires specific solid and liquid media for which the delay is 10 to 20 days. One of the species in this bacterial class, *Mycoplasma pneumoniae*, provokes respiratory infections, most commonly in children or young adults. These infections are often not well recognized due to the lack of rapid reliable diagnostic methods. Bacteriological diagnosis is based on serology, and is difficult to interpret because the time course of the increase in antibody levels is not the same in all patients. Currently the diagnosis can be performed by amplification of species-specific sequences. At this moment, the genome sequence of *M. pneumoniae* is known and available.

5.2.4 *Molecular detection of antibiotic resistance*

Currently, in medical laboratories, antibiotic resistance is determined by the antibiogram method, either in gel medium or in liquid medium. The goal is to detect a
bacterial culture in the presence of an antibiotic gradient which diffuses radially in
a nutrient gel from a disc loaded with antibiotic (gel diffusion method), or in the
presence of a given concentration of antibiotic in a nutrient soup (liquid medium
method). If the culture is positive in the presence of critical concentrations of the
studied antibiotic, the strain is called resistant. These techniques are simple, easy
to set up, with a satisfactory response time and remain the most used. Indeed, the
direct detection of resistance genes in the original samples is a rapid method but does
not have a single interpretation. The samples are often contaminated with bacterial
flora from areas adjacent to the infection site. In these samples, the isolated bacteria are not all of clinical interest and can carry resistance genes. The detection of a
resistance gene can therefore be falsely associated with the strain responsible for the
infection. On the other hand, there exists a wide variety of mechanisms and genes
for resistance (mutations in the genes coding for the target of an antibiotic or in
regulator genes, acquisition of new genes). This genetic diversity makes for a complex choice of probes to test all implicated genes and to define a test having a chance
of being exhaustive. For these reasons, the standard antibiogram remains the most
frequently used method of detecting resistance. The problem comes from resistance
mechanisms that are difficult to detect in standard conditions and so clinical and
epidemiological information are important. Examples would be the detection of me-
thicillin resistance in staphylococci and certain vancomycin resistance phenotypes in
enterococci that can pass undetected through a standard antibiogram. In these two
cases, an additional detection test by PCR or using a marked probe, the resistance
genes absent in the sensitive wild-type strains allow classifying the strains as resistant.
This detection allows adjustment of the antibiotic treatment and the setting up of
isolation procedures to avoid dissemination of the infection from the carriers. With
slow growing bacteria such as *M. tuberculosis*, the detection of rifampicin using PCR,
sequencing, or detection using probes, allows very early adaptation of the treatment
whereas the delay for a response would be very long with classical techniques.

5.2.5 *Bacterial typing*

In communal situations like a hospital environment, the knowledge of reservoirs of
pathogenic organisms and their transmission pathways is important for understanding epidemics, the mechanisms implicated in the occurrence of hospital-acquired
infections and in the dissemination of multi-resistant bacteria, which currently
constitute a major public health problem. During its dissemination, a bacterium
does not stop to divide more or less rapidly. The bacteria coming from these divisions (clones), and detected in several different individuals or in the environment,

determine the chain of transmission. The possibility to determine, in a bacterial species, the strains that form a part of the same chain of transmission from those that are independent greatly helps the interpretation of data from epidemiological studies. As well as sorting strains according to criteria that can differentiate within a species, typing can also determine the links of clonality between isolated strains in a particular epidemiological context.

Historically, typing was performed by the study of phenotypic characters like antibiotic resistance, sensitivity to bacteriophages, activity with respect to various substrates, etc.. However, the lack of reproducibility of the results, their non-quantitative aspect, the often large number of strains which could not be typed and the complexity of certain techniques have favoured the development of molecular typing. Numerous methods and techniques have been proposed exploring different bacterial components such as proteins, isoenzymes, plasmid DNA or total DNA. They reflect the difficulty, whatever the method, of determining the links of clonality between strains. The techniques currently most used to test clonal links between strains are those that analyse the bacterial genome.

Analysis of plasmid content

Bacteria can contain autonomously replicating extra-chromosomal double-stranded DNA in the cell, called plasmids. The study of plasmid content from a bacterial strain is based on simple and cheap methods. After extraction of the plasmid DNA, migration on agarose gel allows the determination of the number and size of the plasmids. The profiles obtained after enzymatic digestion and electrophoresis of plasmid DNA allow comparison of plasmid content between studied strains with a high degree of precision. However, the instability of plasmid content, frequent genetic rearrangements, absence of plasmids in certain strains or the presence of the same plasmid content in independent strains (not having an epidemiological link so a priori without a clonal link) are major inconveniences with the use of this tool as an epidemiological marker.

Analysis of cellular DNA

Restriction polymorphism of total DNA, obtained after digestion of DNA by restriction endonucleases having frequent cut sites and after electrophoretic migration, has been used to compare bacterial strains and evaluate their clonal links. The separated fragments of DNA have sizes from 0.5 to 50 kb. However, the profiles obtained generally contain a very large number of bands of which several are overlapping, making the comparison of profiles difficult.

The analysis of total DNA restriction polymorphisms, after Southern blot hybridization with a hot or cold marked probe, has been proposed. Therefore, the number of bands to analyse is reduced to the fragments homozygous for the probe marked sequences. The localization of the fragments that hybridize to the probe varies with the cut sites and reflects the diversity of the genetic code between the studied strains.

Several probes have been studied, including those recognizing DNA coding for the 16S and 23S ribosomal RNAs. The conservation of these genes in the world of eubacteria allows utilization of a universal probe. Nevertheless, the diversity of the probes obtained varies with the species studies and the marker can display weak discriminatory power. The discriminating characteristic of a method is its capacity to differentiate independent or non-linked strains (without clonal links). In contrast, certain profiles are characteristic of a species and can help the diagnosis of a species when the classical methods do not allow strain identification. Other probes have been evaluated, among them those that carry the sequences coding for virulence genes or the repeated insertion sequences like IS6110 in *Mycobasterium tuberculosis* or IS256 in *Staphylococcus aureus*. These last two probes have increased determination power but are reserved, for technical reasons, to reference centres.

Classical electrophoresis in a unidirectional electric field allows the separation of DNA fragments of a size smaller than 50 kb. Other electrophoresis methods allow the separation of fragments from 50 to more than 800 kb, such as those used in the contour homogenous electric field (CHEF) technique, the most frequently used for bacterial typing (Figure 5.2). The profiles obtained with these methods have generally less than 20 bands and are easily compared. The restriction endonucleases used are those that have only a few cut sites (macrorestriction) in the studied species. The principle of the electrophoresis consists of exposing digested DNA to the electric field, which regularly alternates its orientation. In this case the DNA strands have to regularly re-orientate themselves in the sense of the electric field during their migration in an agarose gel. The time required for re-orientation of the DNA strands is proportional to their size and allows their separation. For the DNA extraction, the bacteria must be included previously in the blocks of agarose (insert). These blocks are then submerged in several solutions (lyse solution, washing) to obtain cellular DNA extracted and non-fragmented on the interior of the block during the manipulation. This method has a high power of discrimination, high reproducibility and is applicable to a wide range of organisms after adaptation of the protocols. However, the delay before obtaining the results is relatively long – more than 3 days – and the necessary techniques and equipment are onerous.

Analysis of genomic polymorphism after random amplification

Among the amplification techniques described for typing bacterial strains, the technique of random amplified polymorphic DNA (RAPD) is the most commonly used (Figure 5.3). This method consists of amplifying several DNA fragments with the help of a random sequence primer (some 10s of nucleotides and a GC content greater

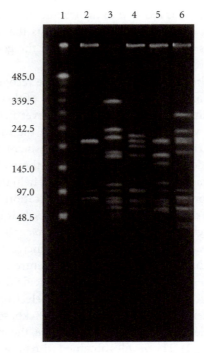

Group B streptococcal strains isolated during an epidemic over the period of 1 month compared by the CHEF technique. All of the profiles are different.

Track 1: lambda ladder (multimers of the phage λ genome, molecular size marker);
Track 2: strain n° 1;
Track 3: strain n° 2;
Track 4: strain n° 3;
Track 5: strain n° 4;
Track 6: strain n° 5.

Figure 5.2   CHEF technique

than 40 per cent) hybridized at low temperatures (between 37 and 45°C) to partially homologous sequences and close enough to each other to allow amplification. After electrophoresis of the amplified fragments, a profile is obtained consisting of a variable number of bands representing the fragments of variable size that characterized the target DNA. This method does not require information on the target sequence. It is rapid, cheap, technically simple to set up and applicable to all bacteria. The essential problem remains with the reproducibility, but this can be improved by rigorously standardizing all steps of the experimental protocol.

Interpretation of results

The bacterial genome is subjected to permanent variation occurring during cell division (random mutations, genetic rearrangements, exchanges of genetic material,

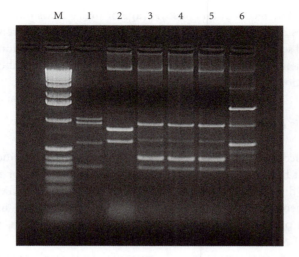

Strains of multiresistant *Klebsiella oxytoca*, isolated from neonatal reanimation during the period of 1 month in six different newborns, compared with the RAPD technique.

Track M: size marker; Track 1: strain n° 1; Track 2: strain n° 2; Track 3: strain n° 3; Track 4: strain n° 4; Track 5: strain n° 5; Track 6: strain n° 6 (identical profiles for strains 3, 4 and 5).

Figure 5.3   RAPD technique

etc.). Genetic variability can therefore be found amongst strains descended from the same parental clone. In theory, this variability is all the more important as the strains become subdivided. Nevertheless, the degree of genetic similarity is generally higher amongst strains recently descended from the same clone than amongst independent strains. For the macrorestriction profiles, it is allowed that strains with profiles differing by less than three bands are strongly related, by four to six bands are potentially related, and by more than six bands are independent or non-related. These criteria should nonetheless be adapted to the studied species thus to the epidemiological context in which the strains were isolated.

5.3   Detection of GMOs

Thanks to the accumulation of scientific knowledge in molecular biology, the first transgenic bacterium was created in 1973, the first transgenic animal, a mouse, in 1981, the first transgenic plant in 1983. A transgenic tomato showing a slow-down in its maturation was put on the market in the United States in 1994 and Europe saw the arrival of genetically modified maize in 1997.

Genetically modified plants arrived at the same moment as the contamination of the blood supply, mad cow disease, dioxine, etc. having already raised the fears of the public. The exercising of their right to free choice led to a European regulation allowing them to be informed of the presence of genetically modified elements in food products. The recent affairs of soya, rape or maize seeds found to be contaminated by traces of genetically modified organisms (GMOs) has renewed the need for reliable and sensitive methods for the detection and identification of GMOs, but their development is still continuing because of the significant number of factors still needing a solution.

5.3.1   What is a GMO?

A GMO is defined as 'an organism whose genetic map has been modified in a different way from what happens in nature by means of cross-breeding or natural genetic combination' (directive CEE 90/220 and French law 92/654). A GMO is therefore a product of genetic engineering[1] that has allowed the transfer, into the genetic inheritance, of one or several genes permitting the modification, suppression or introduction of certain characters.

The introduced genes (transgenes), because of the universality of the genetic code, can come from any organism: virus, bacterium, yeast, fungus, plant or animal. For example, transformation in plants can consist of:

- improving agricultural characteristics (herbicide resistance, resistance to insects, nematodes, fungi, bacteria, viruses, etc.);

- helping the production of 'hybrid variety' seeds between two parental lines: one line produces no more pollen (male sterility), the other allows the restoration of fertility;

- improving the conservation of fruits by delaying ripening;

- improving the nutritional quality of food products by modifying the fatty acid composition of oily grains, by modifying the starch so that it is more digestible for animals and by reducing the allergic properties of a plant (research ongoing in rice);

- production of products for medical use (collagen, gastric lipases, albumin, vaccines, etc.).

GMOs correspond to organisms or parts of living organisms that are biologically active, capable of DNA transfer to other organisms and which can spread into the

---

[1] The introduction of a gene into the genetic inheritance of a plant is carried out in one of two ways as follows. (a) The first method transfers DNA with the help of bacteria from the genus *Agrobacterium* – these soil bacteria naturally transfer a part of their DNA into the cells of certain plants and therefore provoke cell proliferation (e.g. the appearance of galls or root nodules). The genes responsible for the disease in the bacteria are replaced by the genes which are transferred. (b) The second method transforms protoplasts by injecting naked DNA by electroporation or using a DNA gun.

environment; for example in the case of a plant, it is as much a question of the whole plant as of the fruit, grain, pollen, etc. In contrast, derived products such as flour, oil, etc., are not considered as GMOs.

5.3.2 Regulations

Legal context and evaluation techniques

The analysis of health and environmental risks is the fundamental element and a prerequisite before any GMO can be put on to the market (directive CEE 90/220, French law 92/654, CE rule 258/97).

The evaluation of a GMO is aimed at an individual genetic modification or a 'transformation event', that is to say, a genetic construct inserted at a particular place in the genome[2].

In France, the risk analysis linked to the dissemination of a GMO into the environment is performed by the CGB[3] for environmental questions and by the AFSSA[4] for food-related questions (human or animal feed). The evaluation is performed on a case-by-case basis and takes account of numerous elements such as the nature of the transgene, the genetic construct, the species and variety, and the usage to which it will be put (food, industrial or other).

Imported or cultivated GMOs

The voluntary distribution of GMOs in the environment is governed by the directive CE 89/220; the use of GMOs – and their derived products – in foods is covered by the directive CE 258/97. Certain varieties of maize resistant to pyral (a parasite) and/or herbicide tolerant are authorized for culture and/or for importation for their industrial transformation. For transgenic soya, only the importation is authorized with the goal of industrial transformation. One tobacco variety and two chicory varieties are authorized for culture only for producing seeds for export. Carnations modified to last longer are also authorized, but only for the production of cut flowers. Certain GMOs that are authorized in the European Union can be forbidden in a member state, e.g. in France (oil-seed rape moratorium).

Labelling and traceability of GMOs

Rule CE 258/97 states that the labelling must indicate at the level of the finished products destined for human consumption, the presence of GMOs and of their derived

---

[2] If the same genetic construction used for the same species is inserted into a different location, the GMO obtained has to undergo a new evaluation.
[3] CGB stands for the Commission de Génie Bio Moléculaire.
[4] AFSSA stands for the Agence française de sécurité sanitaire des aliments.

products. At the moment, there is not much in the way of GMOs (melon, tomato, etc.) on the European market. In contrast, products derived from GMOs can be used today as food or ingredients (maize flour, maize semolina, maize starch and starch-based binders, derivatives of maize starch (glucose syrup, dextrose, maltodextrins, etc.), soy flour, vegetable fats (maize, soy, rape)). The mentions of labelling (rules CE 1139/98 and CE 49/2000) give some leeway to reply to the problem of fortu-itous 'contamination' with GMOs (involuntary mixture of non-transgenic varieties with GMO varieties by wind dispersal of pollen, technological constraints, etc.). The threshold of tolerance defined by ingredient (for each vegetable species) was fixed at 1 per cent. Food products and ingredients containing additives and flavours derived from GMOs are subject to the same labelling requirements (rule CE 50/2000).

The application of this rule requires the use of two complementary utilities (Bertheau and Diolez, 1999).

- *Traceability from the raw materials*, that is to say the documented trail that allows identification of the origin and nature as well as the destination of the products within a company and at each commercial transaction. In fact, the current EU procedure on labelling relates on the one hand to seeds and on the other to finished products. In contrast, between these two extremes, there is no longer the obligation to mention 'genetically modified'. These gaps create problems for food manufacturers who need to have accurate information about the GM status of the ingredients delivered to them so that they can label the finished products correctly. The traceability of GMOs currently relies on voluntary conformance while waiting for new regulations.

- *Laboratory analyses* to detect and quantify DNA or proteins coming from GMOs (the two targets kept in accordance with the regulations) in the raw materials, ingredients or finished products, analyses that fall under the step of traceability and self checking.

The authorities have at their disposal laboratories to analyse the products as well as classical control methods (document verification, invoices and evaluation of the reliability of the traceability of the professionals).

5.3.3 Detection of GMOs and their derived products

Detecting the production of the modification

It is normally a question of looking for the protein expressed by the transgene by im-munological and phenotypic tests. These tests are mostly used in regions producing and exporting grains and beans such as North America.

Immunological tests   It is necessary to have antibodies available corresponding to the protein being sought. This is not always possible in the case where a genetic modification does not lead to the synthesis of a new protein, but in contrast acts to reduce or increase the production of a protein already present (such as with the delayed maturation of the tomato) or in the case of a genetic construct leading to the synthesis of a protein produced in a organ different from that being tested.

Immunological tests are attractive because they are simple to perform and cheap. However, the main limitation of these methods is that they can only be applied to the raw materials or to products in the nearly raw state. Indeed, thermal or chemical treatments in particular will alter the conformation of the proteins, even degrade them, and so make this type of analysis impossible.

Immunological tests are mostly performed in the form of a dip-stick (strip of paper carrying antibodies and dipped into the solution to test for the antigen) or the ELISA test.[5] The dip-stick is usable in 10 to 20 minutes in the field or in the laboratory, either plant-by-plant, in the case of GM seed production for example, or by batch (collection of multiple plants or grains) to look for the accidental presence of GMOs in the batches of grain. With these methods at least 0.5 per cent GMOs can be detected.

Phenotypic tests   A part of the AOSCA[6] certification, particularly in the case of soya, is based on the effect of applying herbicide to seedlings issued from seed batches that are not meant to contain GMOs. Phenotypic tests are always restricted to easily controllable phenotypes like herbicide resistance. Moreover, they can only be applied to living material, dependent in this case on the capacity of the seeds to germinate.

The interest of phenotypic and immunological tests must not be underestimated since there will be an operational requirement for reliable traceability of origins and processes. These tests can indeed be used for the certification of pure products at the start of production.

Detection of modification at the DNA level

Amongst all the methods for DNA amplification, the most used technique for the detection of GMOs is PCR. To set up this technique it is necessary to have available:

- a sufficient quantity of DNA[7] of sufficient purity[8] because the PCR reaction is sensitive to the presence of inhibitors co-extracted with the nucleic acids – the extraction and purification protocols are adapted to each type of product;

---

[5] ELISA stands for enzyme linked immunosorbent assay.
[6] AOSCA stands for the Association of Official Seed Certifying Agencies.
[7] The efficiency with which the cell and nuclear membranes are broken down is an element determining DNA accessibility in a sample. Various techniques, using detergents and chelating agents to dissolve the membranes and inhibit nucleases, are then used to free the DNA from the matrix.
[8] Purification is generally performed by fractional precipitation, by adsorption on silica or anion exchange resins.

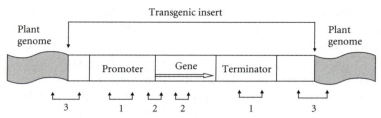

1: Screening P35S/Tnos
2: Identification of the construction
3: Identification of the GMO

Scheme of the insertion of a genetic construct in a plant genome. This construct contains at least the gene of interest surrounded by regulation and transcription signals, promotor and terminator.

Figure 5.4   The PCR targets

- a primer pair (specific oligonucletides) that allows specific amplification of the studied fragment[9] – in order to detect the low numbers of DNA copies, the amplification conditions should be optimized;

- the standard GMO controls to interpret the results – the current calibration standards are based on crushed seeds,[10] but standards in the form of genomic or plasmid DNA could equally well be used.

Qualitative tests   These are based on the capacity to amplify a particular sequence using PCR of a construct inserted into a GMO. Depending on the probe utilized (Figure 5.4) the following results can be obtained (Bertheau, 1998).

- The detection of the presence of GMOs without more information. This test called the 'screen' uses probes in the regulation regions P35S or Tnos (Figure 5.4-1), which are identical in all GMOs authorized in Europe (Lipp *et al.*, 1997);

- The identification of the type of construct present in a GMO by amplification of a part of the transgene (specific probes) – this can be the same in different GMOs (Figure 5.4-2);

- The accurate identification of a GMO by amplification of edge fragments more specific than those of the transgene (Figure 5.4-3)

At the moment, only official laboratories with the task of controlling GMOs have available the edge primers for several GMOs. The modification of directive 90/220

---

[9] In highly refined products, the DNA fragments are less than 400 bp in size, the size of the amplification products should be between 30 and 250 bp.
[10] Reference materials for the quantification of genetically modified organisms (GMOs) issued by the IRMM (Institute for Reference Materials and Measurements).

improves the procedures by which the biotechnology companies make available the edge primers as well as the standards necessary for the analyses. In addition, an important European programme has been put in place to develop quantitative tests based on the edge fragments.

Finally, it is useful to note that the verification of the presence of a non-authorized GMO requires the prior knowledge of the characteristics of the GMO and to have the primers available as well as the control standards. An international register (being set up) of analytical tools would be very useful to perform this type of research.

Quantitative tests   The aim of quantitative tests is to measure the percentage of GMO present in a sample by estimating the quantity of target DNA in the sample using PCR (Larrick, 1997) compared with a range standard. Note that the determination of the GMO percentage in a finished product is defined with respect to each vegetable species. In consequence, except for pure non-transformed products like grains and seeds, the determination of the GMO content of a substrate requires two quantifications.

- The first aims to know the vegetable species contained while determining the number of copies of a DNA target specific for the species. This assumes the availability of a certain number of varieties representative of the worldwide diversity of the considered species, in order to verify the conservation of target sequences and their number of copies.

- The second aims to determine for the same sample the number of copies of a DNA target specific to the location of the GMO for that species. The probes used for the screen (P35S) can only be used for the quantification of pure products (a single vegetable species and even a single GMO because, for example, the number of P35S copies in different GMOs from maize are not the same). Primers specific for the insert can be used in quantitative PCR in a mixture as long as another GMO with the same sequences is not permitted in the EU. Only tests based on edge fragments allow the exact quantification of mixed products, provided that the GMO does not come from gene stacking, that is to say a cross between two previously authorized GMOs.

The ratio of the number of GMO copies of the same vegetable species to the number of specific copies for the species gives the GMO content by species. The final detection variability will therefore be a function of each test PCR, which must be optimized.

To determine the number of GMO copies from a vegetable species and the number of genomic copies of the same species, two approaches can be taken.

- The different types of quantitative PCR known as 'end-point' measures the number of amplicons after PCR by image analysis after gel electrophoresis on agarose in the presence of DNA intercalating agents (for example, BET) or after capillary electrophoresis (see Chapters 1 and 4). For GMO, the method of competitive PCR (see Chapter 1) has been used (Larrick, 1997) but is being replaced progressively by real-time quantitative PCR; it is still used to find out if the GMO content is above or below a certain threshold.

- The different types of PCR known as 'real-time' continuously measure the quantities of amplicon during the PCR (Gachet *et al.*, 1999). Measurement of the increase in amplicon copy number is ensured by measuring the fluorescence liberated by their synthesis thanks to the use of a probe internal to the amplicon (see Chapter 1, TaqMan® chemistry) or by utilization of an agent binding to the DNA of the amplicon (SybrGreen® method).

Figure 5.5 shows the results of a dosage of an unknown quantity of the genome of transgenic maize Bt 176 parallel to a range standard consisting of 100, 1000 and 4000 copies of this genome. The real-time amplification curves (Figure 5.5(a), emitted fluorescence) allow estimation of the Ct (cycle threshold) which, shown on a semi-logarithmic scale (Figure 5.5(b)) with the range standard, permit quantification of the number of genome Bt 176 copies in the tested sample.

This sensitive and specific method requires the use of expensive apparatus.[11] Nevertheless it is used more and more for the quantification of the percentage of GMO in a range of samples.

Statistical approach to the sampling design   This approach studied under the aegis of the FIS[12] and the ISTA[13] is mainly used in the United States for seed exporters. Such an approach allows the acceptance or rejection of a batch of seeds as a function of a fixed tolerance threshold. This method uses qualitative tests (ELISA or PCR) on a certain number of samples of a defined size. The sensitivity of the method depends on the sample number and size and the reliability is expressed in terms of the statistical confidence rather than the analytical precision.

Future evolution of the techniques   The increase in the number of GMOs to detect, authorized or not within the EU, raises the question of the evolution of the detection methods. The 'micro-array' systems and 'DNA chips' or 'branched DNA' could supplant PCR, at least for the cases where the DNA content is sufficient. Given the sensitivity of these hybridization-based methods, it is likely that the response will be positive for seeds, grains, beans and other non-transformed products. In contrast,

---

[11] ABI Prism 7700 SDS (*PE Applied Biosystems*), Gene Amp 5700 SDS (*PE Applied Biosystems*), LightCycler (*Roche Diagnostics*), Icycler (*Bio-Rad*), with a cost of € 50 000–100 000 before tax.
[12] Fédération Internationnale des semenciers.
[13] *Internnational Seed Testing Association*

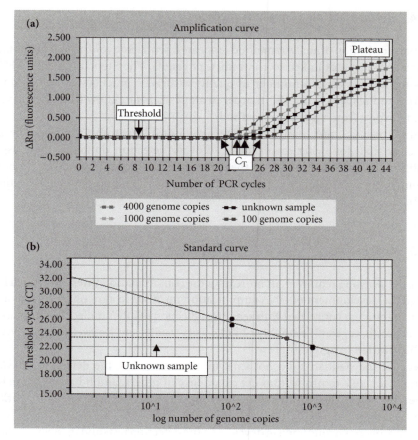

(a) Amplification curve by real-time PCR starting from decreasing quantities of genomic DNA from maize Bt 176. Specific primers for P35S were used.

(b) Standard curve represented by Ct as a function of the log of the initial number of equivalent genome copies of BT 176. Determination of the initial number of equivalent genome copies of Bt 176 of an unknown sample using Ct.

Figure 5.5   Dosage of an unknown quantity of transgenic maize genome

the interest of such hybridization techniques for substrates with low DNA contents, and therefore requiring preliminary amplification before the hybridization, remains to be seen. A recently accepted European programme aims to evaluate the interest of the micro-array techniques both with and without preliminary PCR steps, and to develop the first commercial chip dedicated to GMOs.

The verification of the compatibility of all of these methods must be undertaken urgently in order to reduce as much as possible all chance of discordance between laboratories. The definition of acceptable variability makes up an important factor in the accountability of these methods.

5.3.4   *Harmonization of analysis methods*

The harmonization of analysis methods appears essential to avoid professionals facing commercial disputes linked to contradictory results. The French standard AFNOR aims to define a certain number of quality assurance practices as well as diverse positive and negative controls, in order to allow clients to know what they have the right to expect from an analysis method.

Sampling and under-sampling are recurrent problems to bear in mind, because it is clear that, following the example of other detection domains, these aspects condition the statistical validity of the performed analyses and should be taken into account by the future standard at the level of the CEN[14] (Philipp, Seyler and Bertheau, 2000).

5.3.5   *Conclusions*

It was the labelling requirement more than the contamination threshold which kick-started the development of DNA amplification methods in agriculture and, as a side effect, in the areas concerned with sensitive detection of DNA molecules (human and animal health, vegetable pathology, biodiversity studies, etc.), as well as identification (authentication in agriculture etc.).

Because the standards did not exist in other domains, the detection of GMOs has not been able to profit from the experience of others, and it has therefore been pioneering work. The requirements are a long way from being satisfied but work has begun.

The problems encountered in developing the detection tests and in their use should prompt the proper authorities and the European Commission from now on to only authorize GMOs for which the conditions of reliable detection are met. Free choice would thus benefit in terms of speed, reliability and cost.

5.4   References and Bibliography

5.4.1   *References*

Bertheau, Y. (1998). 'Détection et identification des OGM.' *POUR*, **159**, 69–77.
Bertheau, Y. and Diolez, A. (1999). 'Les aliments passés au crible.' *Biofutur*, **192**, 28–32.
Gachet, E. *et al.* (1999). 'Detection of genetically modified organisms (GMOs) by PCR: a brief review of methodologies available.' *Trends in Food Science & Technology*, **9**, 380–388.
Larrick, J. W. (1997). *The PCR Technique: Quantitative PCR*. Biotechnics Books, Eaton Publishing.

---

[14] CEN stands for the Comité européen de normalisation.

Lipp, M. *et al.* (1997). 'IUPAC collaborative trial study of a method to detect genetically modified soy beans and maize in dried powder.' *J. AOAC Int.,* **82,** 923–928.

Philipp, P., Seyler, J. F. and Bertheau, T. (2000). 'Détection des OGM dans les produits alimentaires et normalization.' *Bulletin des experts chimistes de France.*

5.4.2 Bibliography

Avril, J.-L. and Carlet, J. (1998). *Les Infections nosocomiales et leur prevention.* Ellipses, Paris, France.

Bertheau, Y. and Diolez, A. (2000). 'Détection des OGM: du libre choix des consommateurs aux études de vigilance.' *OCL,*7(4).

Bogard, M. and Lamoril, J. (1999). *Biologie moléculaire en biologie clinique.* Elsevier, Paris, France.

Drouet, E. *et al.* (1997). 'Stratégies moléculaires applicables à l'étude de l'infection virale.' In *Virologie moléculaire médicale* (eds Seigneurin, J.-M and Morand, P.). Tec & Doc Lavoisier (Cachan), Editions médicales internationals, pp. 47–102.

Freney, J. *et al.* (2000). *Précis de bactériologie clinique.* ESKA, Paris, France.

Mazeron, M. C. (1998). 'Diagnostic des infections virales: méthodes virologiques et sérologiques.' In *Traité de microbiologie clinique* (eds Eyquem, A., Alouf, J. and Montagnier, L.). Padoue, Piccin Nuova Libraria editor, pp. 757–764.

Newell, J. O. and Persing, D. H. (1999). 'Applications of molecular amplification methods in diagnostic virology.' In *Laboratory Diagnosis of Viral Infection,* 3rd edn (eds Lennette, E. H. and Smith, T. F.). Marcel Dekker Inc., pp. 111–157.

Tang, Y. W. and Persing, D. H. (1999). 'Molecular detection and identification of microorganisms.' In *Manual of Clinical Microbiology,* 7th edn (eds Murray, P. R. *et al.*). American Society of Microbiology Press, pp. 215–244.

Wiedbrauk, D. L. and Farkas, D. H. (1995). *Molecular Methods for Virus Detection.* Academic Press Inc., San Diego, USA.

6 Identification using genetic fingerprints

Jean-Paul Moisan and Olivier Pascal, *CHU de Nantes*

6.1 Introduction

It was in the 1980s that, for the first time, the idea of DNA variation was highlighted, but it was not until 1985, following Alec Jeffreys' publication, that DNA evidence was accepted in legal cases. It was in fact at that moment that this young researcher (who later would be knighted by the Queen) showed that the DNA of an individual could appear in the form of an image which is unique for each person. Using this tool to identify a murderer who had left sperm at the crime scene, he showed the power of biological evidence as an essential complement to traditional investigation techniques. Since then, a long road has been followed which has brought genetic fingerprinting into the limelight as 'the queen of proofs'.

6.2 Genetic fingerprints by the analysis of nuclear DNA

In the early years, only nuclear DNA was used for genetic identification, and it still remains the most reliable evidence in spite of the undeniable contribution of mitochondrial DNA analysis. Using the comparison between a test sample (whose origin is unknown and that the investigation has the responsibility to identify) found at the scene of a crime and reference samples (DNA of known individuals), the method of genetic fingerprinting depends on three principles: the unique character of the individual genome (only monozygotic or identical twins have strictly the same DNA), the identical composition of DNA in all cells of the human body (allowing the comparison to be made between the DNA contained in sperm with that of blood

---

Diagnostic techniques in genetics   Edited by Jean-Louis Serre

or hair), and the allelic segregation of homologous sequences of genes in the same way as for molecular polymorphisms (allowing for paternity testing).

6.2.1   The molecular and technological basis of the scientific approach

VNTR and microsatellite variable regions

The technology used to reveal genetic fingerprints studies DNA variation using variable number of tandem repeats (VNTR, see Chapter 1) regions. The analysed regions of DNA contain a sequence of bases repeated a variable number of times. The loci studied are located on the autosomes, so the DNA for each individual is characterized (in theory) by a maximum of two alleles. The length of the sequence unit varies, and repeat units of less than six nucleotides are called short tandem repeat (STR) regions. In legal medicine, microsatellite repeats of three (trinucleotides) or four (tetranucleotides) are used the most, whereas the genetic genome map, defined by Genethon, uses repeats of two nucleotides (dinucleotides) which are much harder to interpret. According to the size of the base unit and the number of repetitions, the size of the alleles varies from several tens to thousands of nucleotides. The technique should therefore be adapted to the studied regions.

The choice of a variable region for medical—legal use

It is through use that good products are recognized! Over the last few years, several systems have been evaluated. Some of them are still around, others have disappeared. The criteria of choice are based on the needs of the analysis. For paternity testing, the favoured loci have a low rate of new mutations, whereas for penal applications the favoured regions are short (allowing the generation of genotypes even from partially degraded DNA using PCR amplification), and above all show a high level of inter-individual variation, that is to say a large number of alleles with a population frequency distribution that is as even as possible (see Chapter 7, informativity). Because a VNTR allele is not unique, it can found in the DNA of a large number of individuals, its frequency varies from a fraction of a per cent to more than 25 per cent for the most frequent alleles. This phenomenon explains the requirement of analysing a large number of loci to identify an individual, giving the most specific possible genotype.

  The techniques evolved and STRs, revealing fragments of DNA of several hundreds of nucleotides, are now favoured for all applications. The most commonly used STRs are found in some 20 systems that are known worldwide.

Determination of a genotype relative to a variable region

The old Southern technique (using enzymatic digestion, transfer, hybridization with a marked probe and autoradiography) has been progressively replaced by amplification techniques. Over the period of a few years, PCR has revolutionized genetic fingerprinting with a thousand-fold increase in sensitivity and a reduction in the time for an analysis to several hours (compared with several days using the Southern technique). However, each situation has its advantages and disadvantages: amplification is subject to the risk of contamination (see Chapter 1) and it is necessary to put in place strict quality control procedures to ensure the validity of the results. The extraction step is the most critical and requires particular skill. DNA from a putrefied cadaver is not extracted using the same techniques as those used for a 5 ml tube of blood freshly extracted in sterile conditions. A quantification step is then required to give the ideal quantity of DNA for amplification. Most of the time, PCR is performed using commercial kits, perfectly optimized, which allow characterizing up to 16 loci (VNTR or STR) in the same reaction (Figure 6.1). Such complex analyses are not possible with classical separation and coloration techniques, and only the use of automatic sequencers (on polyacrylamide gels or by capillary electrophoresis) with reading by laser allows differentiation of the co-amplified regions. The transformation of the visual signal into allele sizes requires very specific analysis programs.

Even if many of the steps can be automated, the interpretation is still in the human domain. It is imperative to set rules for the analysis which conform to the manufacturers' recommendations, and which most of the time are simply common sense. How is it possible to allow a reference sample (that is to say of very high quality) to show an inequality in intensity between the alleles when it is simply a matter of regulating the starting quantity of DNA (Figure 6.2)?

Quality control[1]

In addition to the technical performance, the setting up of quality control procedures at each level of responsibility has become essential to the credibility of the analyses.

The experts have an accreditation that lasts for 5 years. This requires that the candidates appear on the list of experts of an appeal court, but they must also have high level degrees (e.g. science doctorates) related to the technology used. In addition, several times a year the French agency in control of the sanitary security of health products uses control samples to evaluate the competence of the experts. All of the internal personnel trained in these specific techniques must follow the written procedures and modify them according to requirements. Keeping up to

---

[1] This section is specific to France.

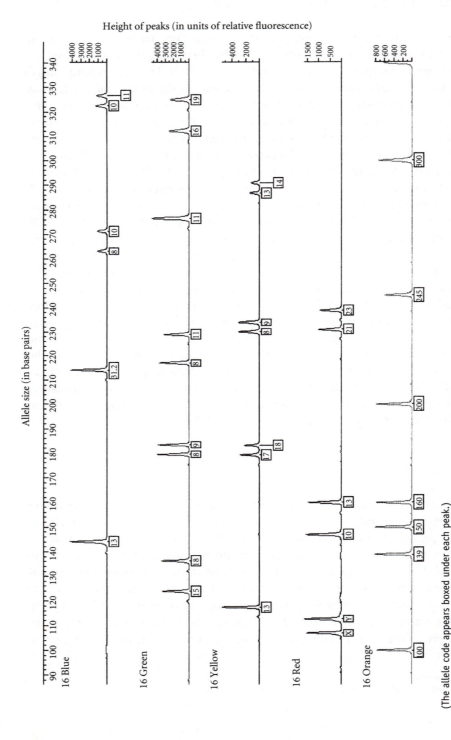

(The allele code appears boxed under each peak.)

Five colours are used, of which orange is for the internal size standard. Several STRs are amplified in each colour with allele sizes that do not overlap. The frequency of a typical profile in the general population is less than 1/10⁹.

Figure 6.1 Genetic fingerprint of a male individual (XY in Amelogenin) defined using 16 loci

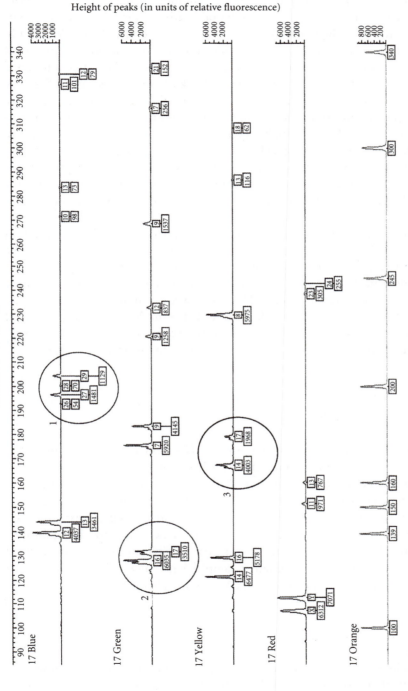

The code and amplification intensity of the alleles appears boxed under each peak.

Note: (1) The appearance of extra peaks called *stutters*: amplification of the allele (*n*-1 base), by replication skid *in vitro*. (2) The shoulder on the leading edge of the principal peak, mostly for small-sized alleles. (3) The disequilibrium of amplification between two alleles at the same locus for a heterozygote individual. (4) General disequilibrium of the genetic fingerprint by preferential amplification of loci having small-sized alleles. The peaks are higher on the left (small-sized alleles) than on the right.

Figure 6.2 Four examples of anomalies appearing during the amplification of too high a quantity of DNA

date with measurement techniques and the maintenance of equipment are both necessary to the validation of the process. In addition to the risk of errors from human activities (tube inversion, switching of reactives), the technique used produces – because of its high sensitivity – a risk of false results. All DNA found in the tube will be amplified, whether it comes from the original sample or not. If the chance of contamination by the investigator at the sampling step is minimal (cheek or skin cells for example), the principal risk lies in the laboratory, firstly by the quantities of DNA handled. If 10 ng are extracted from a cigarette end, there are several μg coming from a blood sample. In the same way, starting from several copies of test DNA, there will be several billion copies of amplicon generated, ready to disappear into the atmosphere in the form of droplets, which can contaminate the neighbouring tubes. In the light of these observations, preventative practices can be implemented: separate extraction of the samples in question (origin to be identified) and the reference samples (known origin), physical separation of the extraction, amplification and sample reading zones, rooms kept under positive pressure, etc.. Controls are also introduced: negative controls for amplification (consisting of all of the reactives for the amplification with the exception of the DNA), negative controls for extraction (tube submitted to all of the analysis steps but containing no biological tissue at the start), and positive controls which are meant to give a known genetic profile. The move towards automating the techniques contributes to reducing all of the risks, as long as the machines are known and are well maintained. Finally, all analytical laboratories accepting the 'Good Laboratory Practices' must in future move towards accreditation to European and world standards.

International nomenclature

At the moment, a uniform analysis system for STRs shown by amplification is being set up. The international nomenclature is based on the number of repetitions comprising the alleles – thus for system TH01, the allele called 6 in the nomenclature is composed of six base units. Moreover, the installation of an automated file in France (and soon in Europe) makes such standardization necessary. During analyses, in order to identify the alleles to be characterized, markers are introduced into the experiments. They are made up of the principal possible alleles, and are analysed in parallel. The comparison of different samples with these references allows the attribution to unknown samples of a code known worldwide (Figure 6.3).

The special case of rare alleles

Some mutations can occur in the middle of a repeat unit causing the appearance of an allele with an unusual size. Consider a system where the repeat unit is composed

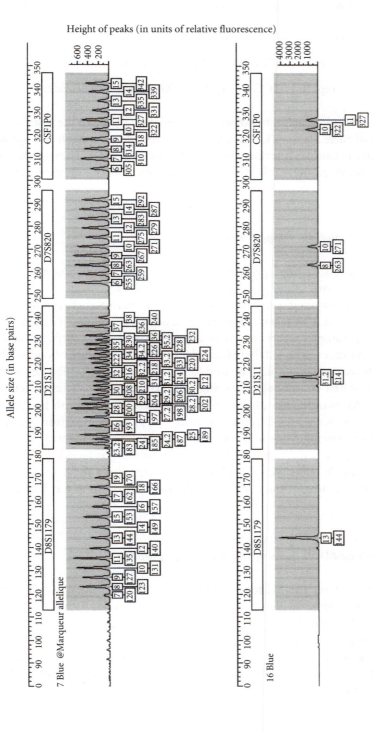

Figure 6.3   Marking

For a given locus (D18S1179, D21S11, D7S820 or CSF1PO), a marker containing the principal alleles (upper line) is analysed at the same time as the samples (lower line). By comparison, the size of the unknown allele can be determined. The precision of the analysis is given as plus or minus one nucleotide. The code and the allele sizes (in base pairs) appear boxed beneath each peak.

of four nucleotides ATCG. The possible alleles are:

- allele 2 (two repeats) ATCG ATCG = 8 nucleotides
- allele 3 (three repeats) ATCG ATCG ATCG = 12 nucleotides
- allele 4 (four repeats) ATCG ATCG ATCG ATCG = 16 nucleotides

Imagine for a given individual, a mutation has caused the deletion of the A and the T in the first unit of allele 3: CG ATCG ATCG = 10 nucleotides. The allele will appear with a size of 10 nucleotides intermediate between alleles 2 and 3. These alleles are extremely interesting, as they indicate in a quasi-formal way the identity of their owner.

Application to the medical–legal domain

In daily practice, a genetic fingerprint must respond to the specific question: who does this biological tissue, found at a crime scene, come from?

To reply to this question, it is necessary to make a comparison between pieces of biological evidence (samples taken at the crime scene and of unknown origin) and reference samples (samples where the origin is perfectly identified: suspect, victim, witnesses, etc.) in order to know if there is an identity or a non-identity between the evidence and the reference sample. The unfolding of the analysis can be described in three steps. The first consists of visualizing and identifying the alleles present in the DNA extracted from the samples. This is an essential step, but can be subject to difficulties or involve errors. During the second step, the comparison is made between the DNA coming from different sources to decide the identity or non-identity of the samples. If the response is positive (an identity is found between an evidence sample and a reference sample), the third step allows the assessment of the strength of the identity of the two samples by the calculation of a risk of error statistic.

6.2.2 The mathematical basis of the scientific approach

All comparisons rest on the existence of two hypotheses where it is necessary to measure the weight of evidence for one against the other. This is why scientists have rapidly introduced, as the first steps of science, tools to measure the weight of evidence. In the medical domain, the best known law is that of Bayes for evaluating the probability of an event conditional on the occurrence of another known event. This law is used directly by some authors, but experts confronted by examinations in their daily routine will use simplified formulae.

The hypotheses of the prosecution and defense

While the exclusion of identity has no risk of error (at minimum two loci differing for nuclear DNA or two different mutations for mitochondrial DNA), the confirmation of identity cannot be so categorical. Indeed, the analysis relates only to the identity of a few loci, which represents for 10 of them, each having a size of 300 nucleotides, around 3000 nucleotides, so just one millionth of the genome. It seems reasonable to imagine that in the general population, two individuals could possess the same genetic profiles (same genotype for the 10 polymorphic markers studied) for this piece of DNA.

There are two opposing hypotheses: that of the prosecution that affirms that the individual has a very small chance of not being responsible for the crime, and that of the defense that espouses the idea that there exists a real chance that the accused is not guilty. The expert, who should not be party to either of the sides, must explain the results in the most even-handed fashion possible, keeping as close as possible to the technical observations. Two hypotheses are compared: the probability that the identified genotype (piece of evidence) is that of the accused over the probability that the identified genotype is not that of the accused.

While the value of the numerator is 1 (the two genotypes are completely identical), the denominator takes into account the frequency of the genetic profile in the general population (that is $1/X$ millions). The result presented by the expert is equal to 1 over $1/X$ and is explained in the following fashion: it is X million times more likely that the unknown DNA is that of the accused than that of an individual taken at random from the general population.

Calculation of the frequency of the genetic profile

Assume that, for a given locus THO1, the DNA of the piece of evidence is characterized by the alleles 7 and 9. The frequency of allele 7 is 18.81 per cent in the general population, and that of allele 9 is 25 per cent, the frequency of the genotype (that is to say the presence of the two alleles in an individual) will be equal to $R1 = 0.25 \times 0.1881 \times 2 = 5.2$ per cent by the Hardy Weinberg hypothesis, see Chapter 7.

The same calculation is repeated for each of the analysed loci, and under the hypothesis that they are in gametic equilibrium (absence of genetic disequilibrium, see Chapter 7), the final frequency of the genetic profile will be equal to the product $R1 \times R2 \times \ldots Rn$. In practice, the use of multiplex systems with more than 13 loci produce frequencies less than 1/one billion, so giving a very low risk of a false identity being made between the evidence sample and a reference sample.

Population studies

These are indispensable since (see above) the calculation of a frequency for a given genotype assumes knowledge of the population allele frequencies. These studies and their results are available for numerous ethnicities. The problem is not so much in the scientific part of the method as in its application in relation to the idea of 'race'. Certain countries are compelled to calculate the frequencies using population studies relating to individuals of the same race as the person accused. It is for that reason that in the USA, three large categories have been defined: Caucasian (white), Afro-Americans (black) and Hispanic (individuals from South America). This division seems a little artificial when the inter-population variances that exist are known, for example in the heart of the South American continent. This step goes against the ethno-geographic observations that completely reject splitting humanity into races suggesting instead a planetary continuum. Aside from that, it is all the more illusory to take account of the differences in marker allele frequencies between populations of different ethnic origin, when the origin of the evidence sample remains unknown and the reference samples come from individuals who come from a mixture of different population origins.

The calculation of the frequency of a genetic profile is performed by reference to the general population. If the enquiry is restricted to the interior of a small group of related individuals (a murder in an Indian village in the middle of the Amazonian forest) or above all to the same family (several brothers or cousins implicated in armed robbery), this calculation is no longer valid. The magistrate must be certain that all other individuals suspected have been tested so that they can be excluded and return the suspect to the general case – that of an individual in the general population.

Significance of the statistical result

The calculation takes into account a statistic based on a comparison in relation to the general population. This is a conservative hypothesis in favour of the suspect because all individuals are taken into account, both males and females of all ages. The population of delinquents is much smaller then this, even smaller because the suspect is not in general taken at random from the general population. Other features of the investigation have given indications (identification by the victim, diverse witnesses, fingerprints, etc.) and the genetic fingerprint is only one part of a puzzle forming the search for the truth.

The search, at random in the general population, for a criminal individual, such as was made in the case of La Plaine-Fougère, is very controversial, the more so because a recent publication has shown that two non-related individuals could have, for nine loci, the same genetic profile.

6.2.3 Applications

Paternity testing

Paternity testing is the area of choice for genetic fingerprinting, since the material to be tested is generally abundant and of high quality. For this application, molecular biology has revolutionized the practice. More than 10 years ago the experts, with the groups and sub-groups of blood, could exclude paternities, but did not dare to affirm them: 'we cannot exclude that Mr Y is the father of child G'. These times have changed. With the exception of a few fossilized dinosaurs, all of the community has switched to the magic of DNA – it is possible to affirm paternity!

An individual's genetic inheritance derives from each of their parents: half of their DNA is received from the mother and half from the biological father. By comparing the maternal DNA with that of the child, it is possible to identify the maternal allele and deduce the allele from the biological father. It is then sufficient to examine the DNA of the presumed father to determine the presence or absence of this allele. For a formal exclusion, the affirmation of paternity is completed by a calculation giving an index and a probability of paternity expressed as a percentage. This is generally superior to 99.999 per cent. Finally, the possibilities of using these techniques for research into maternity or the identification of cadavers by the intermediary of a sister or brother should be noted.

Paternity testing raises ethical questions. It arrives at a painful moment for the families who are tearing themselves apart, and the expert who represents the all-powerful science that makes the split definitive. Does he have the right? Is the true father always the biological parent or rather he who loves the child?

As for the rights of the dead, this has raised many problems concerning a recent show-business affair. Is the right for a child to know stronger than the right of peace for the dead? The proposed amendment to the laws on bioethics takes this element into account by specifying that paternity testing cannot be carried out if the person was opposed to it while he was alive.

Criminology

Rape   It is with sexual aggression that genetic fingerprinting has most helped justice. Before the development of this technique, the biologist depended on the presence of spermatozoa and occasionally managed to determine the blood group. Considering that more than 40 per cent of the population have blood group A and more than 40 per cent have blood group B, the precision of the results can be understood! The preferred form of evidence for the analysis is the swab. The techniques used allow separation of the victim's cells (epithelial cells from the vaginal, anal or buccal regions) from the spermatozoa using the higher resistance of the latter. Demonstration of the victim's genetic fingerprint in the fraction enriched with epithelial cells is an excellent internal control, allowing testing of the experiment at the same time as testing the supposed origin of the sample.

For rapes, aside from the swab, all elements carrying sperm are usable; the slip is an excellent material if it stays dry and is kept out of the light (a similar material allowed characterization, after several years, of the genetic fingerprint of a serial rapist), the slides used by biologists for looking for sperm, tissues, or clothes.

The minimum number of spermatozoa that can be detected is between 100 and 200. This detection threshold has been fixed voluntarily quite high. Reading certain publications shows that some teams work with a single cell. These are set in a particular context where sterile conditions can be maintained from the beginning to the end of the manipulations. In legal medicine, there is no control of the initial operations (sampling from the victim) and numerous contaminants can be introduced. Lowering the threshold of sensitivity requires increasing the number of PCR cycles and multiplies the risk of amplifying foreign DNA present in the initial sample (DNA carried by the investigator or introduced during manipulations in the laboratory), which would result in returning a result without being able to certify its validity.

The condom is an excellent form of evidence provided that the owner leaves it at the crime scene. This object presents an enormous advantage; apart from protecting the victim from all sexually transmissible diseases, it carries a double set of information: inside is found the spermatozoa of the rapist and on the outside it is possible to detect the DNA of the victim. Thus several rapists who pretended to have used the condom with one of their friends, have been confounded by the DNA of the victim present on the outside.

To close this section on sexual aggression, it should be remembered that genetic fingerprinting can only identify the sperm of an individual and confirm a sexual relationship. The notion of consent is still not possible to resolve by these methods!

Other Crimes   All biological traces are analysable providing that the cells are present in sufficient numbers: a tiny blood stain, saliva on a cigarette end, glasses or the necks of bottles, nasal mucus on hoods, samples from car steering wheels. The list is not definitive, and the sensitivity threshold will once again be the determinant. It is necessary to find a balance between maximum sensitivity and the validity of the results.

Hospital applications

Following of a bone-marrow graft   In certain types of leukaemias (cancers of the blood), doctors are lead to grafts of bone-marrow cells. The choice of donor is conditioned by the compatibility of the cells of the receiver and the donor, an important element in graft rejection. The closest individuals are, of course, brothers and sisters or possibly the parents. While this choice of donor is the best with regards to the question of compatibility, it raises the problems of following the intervention. Indeed, after an irradiation that will destroy the blood cell lines of the receiver, the cells of the donor are injected. The practitioners want to be able to diagnose a possible rejection. By the classical systems (HLA in particular) it is very difficult to differentiate between two individuals from the same family (the donor and receiver having been chosen in particular for the compatibility of their HLA systems). Genetic fingerprinting therefore comes to the help of the clinician allowing the DNA of brothers and sisters to be easily distinguished, and to determine if the graft is stable (only donor DNA), if there is a relapse with the presence of chimerism (that is to say a mixture of donor and receiver DNA), or if there a complete rejection of the graft (only receiver DNA).

Genetic fingerprints and prenatal diagnosis   During prenatal diagnosis of genetic diseases (Duchenne's muscular dystrophy, Cystic fibrosis, fragile X syndrome, etc.), some fetal cells (trophoblasts or cells from the amniotic fluid) are sampled by the obstetrician. It is very difficult to visually control the quality of the sampling and to confirm the total absence of maternal tissue amongst the fetal cells. A mixture could lead to an erroneous result and an error in diagnosis. Use of genetic fingerprinting allows the maternal DNA to be distinguished from that of the fetus and the sample purity to be verified. This step has become obligatory in the quality control protocol for prenatal diagnosis.

Genetic fingerprints and twin birth   Diagnosis by genetic fingerprinting of identical or fraternal twins (mono- or dizygotic twins) is important for the case of diseases or malformations *in utero* because it conditions the therapeutic decision of the clinician.

Genetic fingerprints and spontaneous mutations   In the case of a sudden appearance, without family history, of a genetic disease, the biologist will look in the chromosomes or the DNA to find the anomaly responsible for the disease. If this anomaly is not found in the mother, or in the father, it is pronounced to be a *de novo* mutation. It will still be necessary to exclude, by genetic fingerprinting, a false paternity – an event as likely or possibly more likely than that of a spontaneous mutation.

6.2.4   A special case: the Y chromosome sequence

The Y chromosome is a special marker for males. In the same way as mitochondrial DNA is specific for maternal transmissions, the Y chromosome allows following of the paternal line: all boys receive their Y chromosome from the biological father. As with the autosomes, the Y chromosome also possesses variable regions. Unfortunately these show much less variability between individuals or in their alleles.

The analysis of these variable regions can only be considered as a supplement to the variable regions of the autosomes. Their low power of discrimination does not allow them to be used on their own as a proof of identity. However, they can be used in several specific situations.

- In multiple rape, analysis of the variable regions of the Y chromosome allows the number of individuals having had sexual intercourse with the victim to be

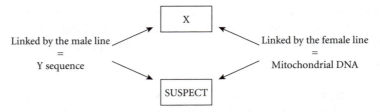

Figure 6.4   Symmetric usage of the Y chromosome and of mitochondrial DNA

counted: the female DNA will not appear, and the DNA from the males will only contain a single allele.

- In the research in a family group where DNA of a suspect individual presents characteristics very close to the DNA of the aggressor it is possible to formally exclude the possibility of identity, but it is likely that the guilty party is a close relative. In such a situation, while mitochondrial DNA analysis allows identification of the maternal line, analysis of the variable regions of the Y chromosome gives information about the paternal line (Figure 6.4). These indications are important to the investigators, as they allow them to limit their investigations to the people designated by the biologist.

- In complex tests of paternity (comparisons with distant relatives), analysis of the variable regions of the Y chromosome brings an essential element in terms of exclusions.

It is indispensable to have in the toolbox the analysis of the variable regions of the Y chromosome, bearing in mind that it is only a complementary technique requiring prudent interpretation.

6.3   Genetic fingerprints with mitochondrial DNA

6.3.1   Introduction

For many years only nuclear DNA was studied, until a researcher, Anderson, published in 1981 the complete sequence of the mitochondrial DNA. This DNA, coming from the organelle responsible for cellular respiration, possesses no link with nuclear DNA. While less discriminating in the analysis of STRs than nuclear DNA, the characterization of mitochondrial DNA represents important progress in identification for legal medicine.

6.3.2   Specific characteristics of mitochondrial DNA

Listed below are the specific characteristics of mitochondrial DNA.

1. The mitochondrial genome is completely different from nuclear DNA. Compared with two times three billion (3×10^9) nucleotides assembled into 23 pairs of linear molecules contained in the nucleus, a mitochondrion has a single circular DNA molecule with just 16 569 nucleotides.

2. In the cell where the nuclear DNA is represented by a unique copy (or two for diploid cells), there exist several tens (or several hundreds) of copies of mitochondria. An enrichment is naturally performed and supports later amplifications.

3. Mitochondria are extremely solid organelles, which can resist extreme conditions that would destroy the nucleus.

4. Mitochondria have variable zones, which are points of mutations at one or several nucleotides. The variation is, however, less informative than observations on nuclear DNA.

5. Mitochondria being transmitted by females, it is simple to identify a cadaver using any relative, provided there is a maternal link (mother, sister, etc.). *A contrario*, unknown DNA found at a crime scene will implicate all individuals in the same maternal lineage (brothers, maternal cousins, etc.).

6.3.3   Utility of mitochondrial DNA

The enormous advantage of nuclear DNA is its large power for discrimination. With current technology, the frequency of a genetic profile in the general population is so low as to render all discussion useless. However, there exist a certain number of cases where nuclear DNA is not accessible, or in such low quantities that it is very difficult to characterize (identification of a body, characterization of excrement or hair).

6.3.4   Methods and techniques

Not all of the mitochondrial DNA is studied; just a region of around 600 nucleotides is sequenced, the control region (Figure 6.5) which presents a sufficiently high spontaneous mutation rate (around one spontaneous mutation every 30 generations), to show inter-individual heterogeneity in the heart of a population. As for the analysis of nuclear DNA, the study of mitochondrial DNA consists of three steps.

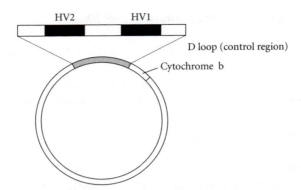

Structure (simplified) of mitochondrial DNA showing the control region with the two variable fragments HV2 and HV1, and the cytochrome b gene.

Figure 6.5   The study of mitochondrial DNA

Definition of mitotype

The sequencing of the control region is carried out. This step is largely automated and only poses problems when the DNA quality is bad or the quantity is very low. As the sequencing step involves the technique of amplification, the risk of contamination problems during the manipulations always has to be kept in mind (see Chapter 1). The comparison of the obtained sequence with the 'Anderson' reference sequence (determined by this author in 1981 from a unique individual) allows the definition of the mitotype, which is constructed from the differences observed between the two sequences.

Comparison of mitotypes

The comparison of mitotypes from a piece of evidence (samples taken from the crime scene) and reference samples (identified individuals) is carried out, in the same way that genetic profiles are compared for nuclear DNA. If the mitotypes from the evidence and the reference sample are different, identity can be excluded. In contrast, if they are absolutely identical, the conclusion of identity has a degree of uncertainty estimated as the error risk.

Determination of the error risk in the event of a conclusion of identity

The characterization of mitochondrial DNA is in its infancy and the problems are more complex than for nuclear DNA. The observed mutations being carried on the same small region, it is dangerous to consider that these polymorphisms are randomly associated as for nuclear markers in gametic equilibrium (see the earlier risk calculation for nuclear DNA and Chapter 7). Obviously most of the mitochondrial markers are statistically associated and the frequency of a haplotype may differ greatly from the product of frequencies of single markers. In this case, it would be scientifically incorrect to calculate the frequency of a mitotype as the product of the frequency of each mutation (or allele), as is done for STRs. This is why the idea of referring to a DNA collection on an international scale prevailed. A bank was set up in the USA, in the laboratory of the American Army in Washington DC, to form the reference. Started with 742 individuals, it is currently composed of 4360 mitotypes from non-related individuals of different ethnic origins (Caucasians, Asians and Africans). The frequency of a given haplotype if given by comparison with this bank, going from 2.6 per cent (the most frequent) to less than 0.02 per cent (no individual found in the bank).

The conclusion of the comparison test and reference samples will therefore be either:

• an exclusion, or

• an affirmation of identity, with as corollary the mitotype frequency in the bank database and above all the note that this mitotype can be found in all individuals in the same maternal lineage.

Several cases exist where this technique of fingerprinting using mitochondrial DNA fails, such as a single nucleotide difference between a cadaver to be identified and their relative. In fact, there is no possible interpretation because it is impossible to know if the two DNAs come from two unrelated persons (a single nucleotide difference occurring by chance) or if the two persons are related and that there has been a new mutation during the transmission of the mitochondrial DNA from one to the other.

6.3.5 Heteroplasmy

Mutations are frequent enough in mitochondrial DNA that, sometimes, the same individual can have two populations of mitochondria (heteroplasmy). At position 16069 (Figure 6.6), the nucleotide described by the Anderson sequence is a C. For an individual Y, a superposition of two nucleotides could appear – a T and a C. Indeed two types of mitochondria coexist in the individual: those where a T is present at position 16069 and those where a C is present in the same individual. Cloning techniques are therefore necessary to be able to separate the two populations. Certain sections of mitochondrial DNA (region 309 and 315) are particularly sensitive to this phenomenon and hair is an extraordinary laboratory for observing the phenomenon.

16 051   16 060   16 069   16 080   16 090   16 100

ACCACCCAAGTATTGACTCACCCATCAACAACCGCTATGTATTTCGTACA
ACCACCCAAGTATTGACTTACCCATCAACAACCGCTATGTATTTCGTACA

The upper line shows the number attributed to each nucleotide, the next line gives the Anderson reference sequence, the last line gives the mitochondrial DNA sequence of F, where a mutation at 16069 replaces a C by a T

Figure 6.6   Anderson reference sequence

Hair is clonal (each hair comes from a single line of cells), whereas blood is polyclonal (all cell lines are represented). For an individual displaying heteroplasmy, one or other of the situations (homoplasmy or heteroplasmy) is determined, according to the analysis of the hair.

6.3.6   Application to the identification of cadavers

Utility of using genetic fingerprints from mitochondrial DNA

Whereas the nuclear DNA of a 'fresh' cadaver is intact, the more time passes the more the phenomenon of cell destruction becomes important. In extreme conditions, the flesh completely disappears and only the bones are accessible. It is sometimes possible to characterize nuclear DNA from old bones (the identification of the Tsar and his family, etc.), but this is a difficult exercise and it is random whether a result is obtained. On the other hand, it is not always possible to have access to close relatives (parents, children) for comparison, while in most cases it is still easy to find a relative from the maternal line (cousins, aunts, etc.). Up until the present day, all cadavers submitted to a study of identification by genetic fingerprinting (the oldest being 200 years old) have been identified using the technique of mitochondrial DNA.

The identification of the remains of the Tsar

At the time of the Russian Revolution, the Tsar, the Tsarina, their children and certain family members were executed by the Bolsheviks, then buried during the night of August 16, 1918. An uncertainty remained concerning the exact localization of the burial and in 1991, a group of Soviet doctors and anthropologists brought to light some human bones from the region of Ekaterinbourg, and it was said that they could be those of the imperial family. An Anglo–Russian group took charge of the study and nine bodies could be reconstituted (four male and five female). The examination of the DNA was immediately considered and the anthropological study and maternal line descendants from the families of the Tsar and of the Tsarina were sought. The analyses showed that the tomb contained the remains of the Tsar, the Tsarina and those of two of their children. Two children were missing, the Tsarevitch and Anastasia. However, for many years a woman in the USA had claimed to be the princess. This supposed Anastasia had unfortunately died several years earlier and had been incinerated. By chance, a fragment of her intestine (taken at the time of a biopsy) had been conserved. The analysis performed by two independent laboratories showed that the mitochondrial DNA was different from that of the Tsarina. The truth was much less romantic than the legend: Anastasia was an imposter. Her identity was definitively established by comparison with an inhabitant of Pomerania who was said to possibly be the nephew of this woman.

The mystery of Louis XVII

As everyone knows, the French Revolution saw the fall of King Louis XVI and of Queen Marie-Antoinette. Their children, the Dauphine and the Dauphin were imprisoned, with the last place of imprisonment being the Temple where they died on June 8, 1795, and were then buried in the cemetery of the church Sainte-Marguerite in Paris. However, most historians claim that the child buried in the cemetery Sainte-Marguerite was not the young Louis XVII. While this point appears to be accepted, the rest remains enigmatic: what happened to the child? Was he buried at the cemetery Sainte-Marguerite? Did he die prematurely and another child substituted to serve as a hostage to the Convention? Could he have escaped from imprisonment?

Shortly after the revolution, around a hundred people claimed to be the true Louis XVII (including a person of mixed race). Most were easily disproved without trouble, but some of them appeared more credible. Amongst them, a certain Naundorff (genial inventor of arms and explosive devices to some, swindler to others) was so credible that he was buried in 1845 at Delft in Holland under the epitaph 'Louis XVII, King of France and Navarre'. A bone coming from the tomb of Naundorff, taken at the time of an exhumation in 1950, could be studied. The comparison was made, after a genealogical study, with the descendants, from the mothers, of the mother of Marie-Antoinette. The two DNAs differed by three mutations, allowing the exclusion of Naundorff as being Louis XVII. The descendants of Naundorff contested the origin of the bone and claimed that a substitution had taken place, the bone not having been kept sealed since 1950. Only a new exhumation and re-sampling would allow a new analysis and certain results.

However, recently a new study was carried out on a heart that could be that of the child autopsied at the Temple. The analysis revealed that the mitochondrial DNA of the heart was identical to that of the descendants of Marie-Antoinette. The conclusion was therefore clear: Louis XVII died at the Temple, without descendants. Unfortunately the mitotype is not unique (0.2 per cent) and above all the traceability of the sample could not be more certain than the bone of Naundorrf. The polemic is all the stronger because the heart of the first Dauphin (elder brother of Louis XVII who died in 1789) had been conserved at the time of the death of the child, but mysteriously disappeared during the 19th century.

6.3.7   Application to the identification of hair

The analysis of mitochondrial DNA allows an extraordinary advance in the identification of hair. This material is largely found in offences and crimes linked to robbery (simple robbery, armed robbery, hostage taking) and terrorism. Hoods, clothes and various objects are repositories for hair. Hair consists of two parts.

- The bulb, responsible for growth of the hair, contains the cells containing both nuclear and mitochondrial DNA. Unfortunately hair found at a crime scene usually has fallen naturally (and up to 50 are lost each day), that is to say that the bulb is dead. The quantity of nuclear DNA extracted is often insufficient to allow characterization by nuclear STRs. Most of the time only mitochondria remain accessible.

- The stem is composed of keratin, but retains a large number of mitochondria.

This hair is the focus of attention of the enquiries. Unfortunately limitations exist: the low quantity of DNA which does not allow the obtaining of results in all cases, the low discriminatory power (high frequency of certain mitotypes), and the possible implication (without possible discrimination) of all individuals belonging to the same maternal lineage.

6.3.8   Applications to the discrimination between man and animal, or between animals

Since the cytochrome b gene, situated in the mitochondrial DNA, shows a specificity linked to the kind of animal, it is possible, by analysis of this region (different from the control region), to determine if a DNA sample is of human origin or, if not, to determine the animal from which it comes. The technique is so powerful that it is possible to distinguish between Pacific and Atlantic salmon. This technique is mainly used in the field of agroalimentation in the study of fraudulent contamination (presence of chicken meat in a product certified as pure beef, presence of pork in food destined for Muslim countries, etc.).

6.3.9   Conclusions

The analysis of mitochondrial DNA, even with a large amount of automation, remains long and costly. The power for discrimination never reaches that of nuclear DNA. Nevertheless the analysis gives important information about elements which would have been impossible to analyse up to now like hair or bones. The interpretation of the results has to be made with extreme care and the limits of the analysis explained to the judge.

6.4   Society facing the question of genetic fingerprint files

6.4.1   Introduction in the form of an anecdote

In 1995, the authors' laboratory worked in collaboration with the criminal investigation unit of Paris on the unsolved murders of several young women. It was certain that they had been performed by the same man, because the same genetic fingerprint was found at the crime scenes. This man was also supposed to have committed other murders (cases J. and F.) but, unfortunately, no biological argument allowed confirmation of this. For one of the victims (case F.), a blood stain was found without knowing if it came from the perpetrator or from a close friend or relative of the victim. It is in this context that the examining magistrate regularly asked the authors to compare the genetic profile of this stain with samples taken from suspects. And thus, in 1995, a suspect X was excluded from being at the origin of the blood stain.

In March 1998, the authors were again working on case F. and this time on the hair. Several unknown DNAs had been characterized leading to re-examination of all of the suspects on file, whose nuclear DNA had therefore been characterized but not their mitochondrial DNA. And this was the breakthrough! When the mitotype of X appeared it was recognized, because since 1995 the mitochondrial DNA of the serial killer had been characterized and had been kept on file. Several days later, X was arrested. Was this a reason for happiness or pride? Without doubt it would have been if two young women had not been murdered in 1997. If a true DNA file had existed in 1995, those two women would still be alive today.

6.4.2   Creation and maintenance of genetic fingerprint files

In the USA and the UK

It is in the US and the UK that the first genetic fingerprint files appeared. The British took the lead in characterizing all individuals who had committed a crime or an offence – but at what price! An enormous intellectual and financial investment, but giving a file with extraordinary power. The database is maintained by the Forensic

Science Service, a private company under contract to the British government. In the rest of Europe, the governments are more cautious and few have taken this step. The affair of the serial killer of East Paris was the catalyst which, in France, brought the problem to people's attention.

The French situation

On the technical level: operation of the file   On the technical level, anarchy presided over the setting up of genetic fingerprinting in France, notably with the difficulty in standardizing the markers. However, to be able to compare, it is necessary that all laboratories characterize the same variable regions of DNA. A commission made up of experts decided in favour of a European model (seven DNA regions to analyse). This decision is even more important given that the techniques are largely evolutionary and it can already be seen that DNA chips are revolutionizing DNA analysis in the same way that DNA amplification replaced the Southern technique in less than 10 years. The chosen system will define for several years the characterization of DNA for judicial uses. Of course, nothing stops working in duplicate, using at the same time the old techniques for the file and the new for efficiency and speed. Note also that the choice of informatics system for management will strongly influence the efficacy of the database.

As this file essentially concerns sexual offences, a system will be introduced to conserve on the one hand unknown DNAs (traces) and on the other hand DNAs from individuals definitively condemned (references). The DNA of suspected persons will be able to be compared to the file but in no case will their characteristics be entered into the bank. The demands will be overseen by the magistrates in charge of the files. The files will be installed in the heart of the Interior Ministry and the exhibits kept by the soldiers of the Gendarmerie.

On the ethical level: the dangers   There is no attack on freedom as long as genetic fingerprints are used with a judicial aim, in a democratic country. No one cries scandal when digital fingerprints are taken to obtain an identity card. Today, only criminals could fear leaving their genetic fingerprints in a database.

The real danger comes from the analysis of DNA for other purposes than judicial. For police usage, the fragments of characterized DNAs are anonymous, their function in the body being unknown. However, DNA contains coding or regulatory regions that are known to have important effects on the functioning of the organism, whether the anomalies lead to genetic diseases, or if they lead to a predisposition to some multifactorial pathologies. It is in this domain that the danger is strongest. An insurance company could multiply the premium if they could prove that a client carried a susceptibility gene for breast cancer, or an employer could refuse to hire a man under the pretext that he carries a gene increasing by 10 the risk of a heart attack. Does not the risk come from a misuse of DNA rather than judicial use intended to protect citizens?

Current state and perspectives While the law in France for the protection of minors, which contains the birth certificate on file, dates from 1998, the structures allowing comparison are not always operational, and probably will not be before the year 2003. The reality of such a tool can only be imagined with the usual material and people given to investigators (police and gendarmes) and to magistrates to make things work. In parallel with the new laws protecting individual liberties, it is time to become aware that the technical and scientific police force represents one of the axes of the development of justice in the 21st century.

6.5 Conclusions

Has the system moved from consensual proof to scientific proof? Is genetic finger-printing the 'queen of proofs'? In the authors' opinions, certainly not. The scientist or the expert holds a part of the truth. They are capable of identifying the owner of a biological trace, and their contribution stops there. It is not up to them to demonstrate the guilt of an individual. This role is held by the jury of the court who listen to all of the elements assembled by the presiding judge over the course of the enquiry. The DNA evidence is only one of the pieces of an immense puzzle and to forget this would give to the scientists and experts a power to which they do not have the right. Let us use and abuse the techniques of scientific investigation, but keep them in their context. Thus the identification of sperm from a sample does not automatically make its owner a rapist, as the absence does not prove innocence. The discovery of DNA of its polymorphisms has advanced genetics by giant steps over the last half century. The domain of justice has greatly benefited from genetic fingerprinting which is, without being the 'queen of proofs', an enormous help for the protection of society in the search for truth while minimizing the risk of errors.

6.6 Bibliography

Cabal, C. (2001). 'Le valeur scientifique de l'utilisation des empreintes génétiques dans le domaine judiciaire.' Rapport de l'office parlementaire d'évaluation des choix scientifiques et technologiques, 5 juin 2001, no. 3121, Assemblée nationale.

Gill, P. *et al.* (1994). 'Identification of the remains of the Romanov family by DNA analysis.' Nature Genet., **6**, 130–135.

Jeffreys, A., Wilson, V. and Thein, S. L. (1985). 'Individual-specific fingerprints of human DNA.' Nature, **316**, 76–79.

Jehaes, E. *et al.* (1998). 'Mitochondrial analysis of remains of a putative son of Louis XVI, king of France, and Marie-Antoinette.' Eur. J. Human. Genet., **6**, 383–395.

Piercy, R. *et al.* (1993). 'The application of mitochondrial DNA typing to the study of white caucasian genetic identification.' Int. J. Leg. Med., **106**, 85–90.

7 Molecular genetics and populations

Jean-Louis Serre, *Université de Versailles*

7.1 Hardy—Weinberg equilibrium and measures of genetic diversity

For all genes, the genetic diversity of a population can be defined and measured at the phenotypic, genotypic and allelic levels, by the frequencies of the various pheno-types, genotypes and alleles that are observed in the population. These frequencies – called phenotypic, genotypic and allelic, respectively – are directly measurable when studying codominant phenotypes, where each phenotype corresponds to a single genotype. In this case, the genotypic frequencies are equal to the phenotypic fre-quencies, and the frequency of each allele of the gene is equal to the frequency of the homozygote plus half the frequency of the heterozygote carriers of the allele. This situation is currently found with polymorphic DNA markers (RFLP, microsatellites, SNP or VNTR).

However, for a gene with certain alleles having dominant or recessive effects, the genotypic and allelic frequencies are no longer directly measurable. In this case, it is currently necessary to use the Hardy–Weinberg model to estimate the allelic and genotypic frequencies (Box 7.1).

7.1.1 Analysis of recessive diseases

Applied to recessive autosomal diseases, the Hardy–Weinberg model (Box 7.1) allows calculation of the frequencies either of the disease alleles, healthy carriers, couples at risk or all of the parameters useful for an epidemiological analysis or calculation of risk (Table 7.1)

Diagnostic techniques in genetics   Edited by Jean-Louis Serre
© 2006 John Wiley & Sons, Ltd

Box 7.1   Hardy–Weinberg equilibrium

The factors which can modify the genetic composition of a population are:

- mutation, migration and selection,

- sampling variation due to small population size,

- patterns in the choice of spouses, that is to say deviations from panmixia (non-random mating).

The theoretical model of 'Hardy–Weinberg equilibrium' corresponds to a population respecting the three conditions of panmixia, large population size and the absence of mutation, migration and selection. Under these three conditions, the genetic composition of the population is invariant:

- the allelic and genotypic frequencies remain constant from generation to generation,

- a relationship is established between the allele frequencies and the genotype frequencies – the genotype frequencies are simply the product of the frequencies of the relevant alleles.

This relationship between the allele and genotype frequencies, called panmictic equilibrium or Hardy–Weinberg equilibrium, is very useful for estimating the allele frequencies of genes controlling recessive traits, notably the allele frequencies of disease-causing alleles responsible for genetic disorders, and the frequency of healthy carriers.

For a diallelic gene with alleles A and a, recessive with respect to A, and with allele frequencies p and q respectively, the following table can be created.

Genotypes	A/A	A/a	a/a
Phenotypes	[A]	[A]	[a]
Phenotype frequencies	$f[A]$	$f[A]$	$f[a]$
Genotype frequencies under H.–W. equilibrium	p^2	$2pq$	q^2
Estimation of allele frequencies under H.–W. equilibrium	$p = 1-q$	$p = 1-q$	$q = \sqrt{f(a)}$

Table 7.1   Application of the Hardy–Weinberg model to several recessive autosomal diseases

Disease	Population	Observed frequency at birth: $f[a]$	Calculated frequency of 'the mutation': $q = \sqrt{f[a]}$	Frequency of healthy carriers: $2q(1-q)$	Ratio of healthy carriers to affecteds: $2(1-q)/q$
Cystic fibrosis	France	1/2500	2%	4%	100
Phenylketonuria	France	1/16 000	0.8%	1.6%	250
Galactosemia	France	1/40 000	0.5%	1%	400
21-OH deficiency	France	1/10 000	1%	2%	200
11β-OH deficiency	France	1/100 000	0.32%	0.63%	625
Sickle cell anaemia	France (Antilles)	1/400	4.7%	9.5%	40
Sickle cell anaemia	Africa (Great Lakes)	4/100	20%	32%	48
β−thalassaemia	Mediterranean	1/100	10%	18%	18

For most heritable diseases that affect lifespan or fertility, that is to say fecundity, the condition of absence of selection on the Hardy–Weinberg equilibrium is not strictly valid, but it is easy to demonstrate that the effect of selection on several generations is sufficiently weak that it can be ignored in the calculation of genotypic and allelic frequencies. Moreover, the frequency of a disease and that of the mutation responsible have very often stabilized over time because the effect of selection is counterbalanced by that of another factor (see below).

The heterozygote frequency can appear quite high when looking at a fairly rare disease (for example, cystic fibrosis with one healthy carrier in 25 in the general population), but for most recessive diseases, the great majority of mutated alleles present in the population are carried by the heterozygotes. This fact is illustrated by the value of the ratio $2pq/q^2 = 2p/q = 2(1-q)/q$ (last column, Table 7.1) that expresses the number of healthy carriers, heterozygotic for a disease mutation, in terms of the number of affected individuals, carriers of two disease alleles. This ratio is higher the rarer the allele.

Note 1   With clinical analysis, it is not possible to distinguish between the different disease mutations: the frequency q therefore corresponds to the frequency of all of these, collected together under the title 'disease mutation'. These 'homozygotes' are revealed, at the molecular level, to be compound heterozygotes (carriers of two different disease mutations).

Note 2   For the great majority of known recessive diseases (about 1500), the mutations are so rare that q^2 is essentially zero and the few examples of the mutation are carried by heterozygotes. In this case, the few observed homozygote patients result mainly from marriages between relatives and not from panmictic marriages.

Table 7.2 Application of the Hardy–Weinberg model to several autosomal dominant diseases

Disease	Population	Observed frequency at birth: (F)	Calculated frequency of the mutation: $p = 1 - \sqrt{(1 - F)}$	Number of heterozygotes for one homozygote: $2(1 - p)/p$
Familial hypercholesterolemia	Europe	1/500	10^{-3}	2000
Neurobibromatosis	Europe	1/3000	17×10^{-5}	12 000
Myotonic dystrophy	Europe	1/7000	7×10^{-5}	28 000
Huntington's disease	France	1/10 000	5×10^{-5}	40 000
Achondroplasia	Europe	1/20 000	2.5×10^{-5}	80 000
Marfan syndrome	Europe	1/25 000	2×10^{-5}	100 000
Osteogenesis imperfecta	Europe	1/25 000	2×10^{-5}	100 000
	Denmark	1/4857	11×10^{-5}	18 346
Retinoblastoma	Europe	1/30 000	1.7×10^{-5}	120 000

7.1.2 Analysis of dominant diseases

The use of the Hardy–Weinberg model is also useful for the calculation of allele frequencies for genes with dominant-acting mutations, and notably for genes responsible for dominant genetic diseases. In the latter case, the healthy phenotype is without ambiguity a/a, while the affected phenotype is A/a or A/A (if this one is not lethal). Under the hypothesis of panmixia, the frequency F of affected individuals is then equal to $p^2 + 2pq$, which means that the complementary frequency $(1-F)$ of the healthy individuals is equal to q^2.

Knowing F, the frequency of affected individuals, we can directly infer the frequency of the functional allele by $q = \sqrt{(1-F)}$, or the frequency of the mutation responsible by:

$$p = 1 - \sqrt{(1 - F)}$$

since $p = 1-q$.

We can consider, if the disease is rare as is often the case, that the homozygotes are very rare (Table 7.2, last column), or missing if they are lethal. All the affected individuals are then heterozygotes and the frequency of the mutation can be directly estimated as being equal to half the frequency of the affected heterozygotes:

$$p = F/2$$

It is interesting to note that dominant diseases, although having frequencies similar to those of recessive diseases (at the phenotypic level), result from mutations that are much rarer at the genotypic level (1000 to 10 000 times rarer). This comes from the fact that individuals affected with a recessive disease are all homozygotes, whereas individuals affected with a dominant disease comprise both homozygotes and above all heterozygotes, which are always more frequent. However, selection – which has no effect on the healthy carriers of a recessive mutation – exerts an effect equally on

Table 7.3 Allelic, genotypic and phenotypic frequencies for sex-linked diseases (X chomosome) (note that p is the frequency of the dominant allele, and q that of the recessive allele)

Phenotypes	Male sex		Female sex		Result: frequency of the disease in each sex
	Affected	Unaffected	Affected	Unaffected	
Genotype frequency (recessive disease)	q	p	q^2	$p^2 + 2pq$	$q^2 < q$
Genotype frequency (dominant disease)	p	q	$p^2 + 2pq$	q^2	$p^2 + 2pq > p$

homozygotes and heterozygotes for a mutation responsible for a dominant disease, which limits the allele frequency. In this case, the heterozygote/homozygote ratio is much higher with dominant diseases and gives further justification for assuming that, without exception, homozygotes are rare enough to assume that affected individuals carry a single disease-causing allele.

7.1.3 Analysis of sex-linked diseases

In the case of a gene located on the X chromosome and controlling a 'sex-linked trait', the situation is appreciably different. In humans (or drosophila), males are hemizygotic for all genes carried by the heterochromosome X. Whether a gene mutation has a dominant or recessive effect with respect to the wild-type allele does not matter in this case, and all males having a given phenotype will be carriers of the allele associated with the phenotype, which allows a direct estimation of the allele frequency from the phenotypic frequency (Table 7.3, columns 2 and 3). In contrast, the genotypic and phenotypic frequencies in females are very different from those in males, even under Hardy–Weinberg equilibrium (Table 7.3, columns 4 and 5) when we assume the allele frequencies are equal in the two sexes (they will become so after several generations if, after a mixture of populations, they were not already).

This analysis allows us to predict that a sex-linked disease or trait will become (under Hardy–Weinberg equilibrium) more frequent in males if it is recessive, and more frequent in females if it is dominant. This prediction is clearly seen for recessive diseases (Duchenne muscular dystrophy, haemophilia), but not for dominant diseases, as the expected excess in females is statistically negligible. In addition, this excess is counterbalanced by biological phenomena linked to X inactivation in the cells of females.

7.2 Multiple alleles and informativity

The Hardy–Weinberg model is not only applicable to the diallelic situation with monofactorial diseases or RFLP markers, but is also very useful for multiallelic markers, notably when molecular analysis of DNA allows distinguishing either the different disease alleles of a gene, or the different alleles of a VNTR marker or microsatellite. In this case, with a multiallelic marker or gene, we can distinguish n allelic types named $A_1, A_2, \ldots, A_n$ with allele frequencies $p_1, p_2, \ldots, p_n$.

Box 7.2   The number of possible genotypes for a multiallelic gene or marker

Under the conditions of the Hardy–Weinberg model, in the absence of mutation, selection and migration and in a population of infinite size, the allele frequencies are stable; under panmixia, the system of randomly selecting two gametes from an urn allows the construction of the next generation. It is simple to see that the table with four cells obtained in the case of a diallelic gene is replaced by a table with n^2 cells in the case of a gene or marker having n different alleles.

Gametes	A_1	A_2	A_i	A_n
(frequency)	p_1	p_2	p_i	p_n
A_1	A_1/A_1		A_1/A_i	
p_1	p_1^2		$p_1 p_i$	
A_2		A_2/A_2		
p_2		p_2^2		
A_i	A_i/A_1		A_i/A_i	
p_i	$p_i p_1$		p_i^2	
A_n				A_n/A_n
p_n				p_n^2

We can see that the homozygotes correspond to the genotypes on the leading diagonal and that there are as many as there are different alleles. The heterozygotes then correspond to the rest of the table, but since we should not count the same genotype twice, we must divide by two the number of cells remaining after removing the leading diagonal. We thus obtain the number of heterozygotes from the n^2 cells in total in the table minus the n homozygotes, all divided by 2, that is to say:

$$(n^2 - n)/2 = n(n - 1)/2$$

The total number of genotypes is equal to:

$$n + n(n - 1)/2 = n(n + 1)/2$$

If there are 13 possible alleles for a marker (for example, a microsatellite, see Chapters 1 and 6), the population will have 91 different genotypes.

For a diallelic gene or marker we count two homozygotes and one heterozygote. For a gene with n distinct allelic types, we count n distinct homozygotes and $n(n-1)/2$ heterozygotes, which makes in total $n(n+1)/2$ distinct genotypes, and as many phenotypes, if there is codominance, notably for polymorphic markers (Box 7.2).

The genotypic frequencies of the n homozygotes are equal to the square of the frequency of the allele they carry, and the frequencies of the $n(n-1)/2$ heterozygotes are equal to twice the product of the frequencies of the two alleles they carry.

In the case of a multiallelic gene, the Hardy–Weinberg relationship between the allele and genotype frequencies for a diallelic gene,

$$(p+q)^2 = p^2 + 2pq + q^2$$

is generalized to give:

$$(p_1 + p_2, + \ldots, +p_n)^2 = \Sigma p_i^2 + \Sigma 2 p_i p_j \text{where} (i > j)$$

Note: The heterozygosity H defined as the frequency of heterozygotes in the population, is equal to $1 - \Sigma p_i^2$.

Insofar as many genetic analyses (linkage mapping of a gene or genetic fingerprinting) take the heterozygosity as being 'informative', the value of H is often described as 'the informativity' of a gene or a marker in a population. We can see that the maximum value of H is never higher than 50 per cent for a diallelic gene or RFLP marker, where the frequencies p and q are equal (1/2), whereas it can easily reach 80 or 90 per cent with a multiallelic gene (HLA) or a microsatellite. If the n alleles are equally frequent, the value of the allele frequencies will be equal to $1/n$ and the maximum value of the heterozygosity will be equal to:

$$H_{\max} = 1 - \Sigma(1/n)^2 = 1 - n(1/n)^2 = 1 - 1/n = (n-1)/n$$

7.3 Selection–mutation balance and Haldane's rule

Disease alleles, even if they are deleterious, are never totally eliminated if we consider that at each generation a certain number of *de novo* mutations arrive which partially replace the mutations lost through natural selection.

It is simple to think that an equilibrium of the frequencies of disease alleles will be established when the flow of alleles eliminated by selection is counterbalanced by an equal flow of new mutations.

In the case of a recessive disease, it has been shown that the equilibrium frequency of 'the disease mutation' is equal to $q = \sqrt{(\mu/s)}$, where μ is the rate of *de novo* mutations and s the selection coefficient (percentage of fecundity loss, equal to 100 per cent in the case of lethality before reproductive age). In the case of a dominant mutation, the equilibrium frequency of 'the disease mutation' is equal to $p = \mu/s$. In the case of sex-linked diseases, the situation was analysed by the geneticist Haldane according to a rule which bears his name and which is demonstrated in a simplified form below.

Since women are always healthy carriers and are never affected (see above), we can consider that selection only acts upon males where, because they are all hemizygous, all carriers are subjected to the effect of selection. All of this occurs as if selection was restricted to males where, because of hemizygosity, the effect of the mutation is as if it was dominant. Since males contain only a third of the mutations in the population (assuming a sex ratio of 50 per cent), only this third of mutations is subjected to selection, so we can write the equilibrium frequency as:

$$q = (\mu/s)/3$$

In the case of lethal diseases such as Duchenne muscular dystrophy, the value of s is equal to 1 and the equilibrium value is equal to $\mu/3$, which signifies that a third of affected boys are carriers of a new mutation. In consequence, a third of mothers of affected boys are not carriers, so preventative DNA testing will not work for these cases.

7.4   Diagnosis with genetic testing: cystic fibrosis—an academic case

Cystic fibrosis is a severe childhood Mendelian recessive disease (monogenic), most frequent in Europe with one affected child in 2500 births. In the absence of non-palliative therapies, many at-risk couples consult a genetic counselling service for prenatal diagnosis (PND). However, diagnosis is only an option if the couple know that they are at risk, most commonly because they already have an affected child, which allows PND for subsequent pregnancies (Figure 7.1). PND can nowadays be successfully performed either directly by identification of the mutations, or indirectly using markers (see Chapters 1 and 2).

The practice of PND over the last 15 years brings couples who are the descendants or relatives of couples at risk to consider a PND for the first pregnancy, where their risk, without being equal to 1/4, is considerably higher than that for a random couple. However, the use of PND requires that certain conditions should be met (Box 7.3).

Box 7.3   The risk of having a child affected with cystic fibrosis for various types of couples

The risk for a couple who already have one affected child (II-1 and II-2) is 1/4 and a PND can be proposed for subsequent pregnancies. The risk for a couple from the general population without a family history (II-5 and II-6) is 1/2500.

The risk for the couple (II-3 and II-4) of having an affected child depends on two events.

- Is II-3 a healthy carrier? The risk of this event, $R1$, is increased because his sister is one.

- Is II-4 also a healthy carrier? The risk of this event, $R2$, is known – it is equal to the frequency of healthy carriers in the general population, so 1/25 (see Table 7.1).

The risk of an affected child is equal to $R1 \times R2 \times 1/4$, so **R1/100**. What is, therefore, the value of $R1$?

If II-2 is a healthy carrier, then she received a mutated allele from one of her parents. It is known, therefore, that one of the parents of II-2 is N/m; the second can also be N/m with frequency 1/25 (the frequency of heterozygotes in the general population), or N/N with the complementary probability 24/25.

In the first case, the probability that II-3 is a heterozygote is 1/2 if his phenotype is not known, but knowing that he is not affected this probability is 2/3. In the second case, the probability that II-3 is a heterozygote is 1/2. The global risk $R1$ that II-3 is a heterozygote is $(1/25 \times 2/3) + (24/25 \times 1/2) = 38/75$. The risk of an affected child being born is then $R1 \times R2 \times 1/4 = 38/75 \times 1/24 \times 1/4$ or approximately 1/197. This is 12.5 times greater than the risk in the general population (1/2500).

Very often, the mutation found in II-2 has been identified during a genetic study using PND (or indirectly identified using a polymorphic DNA marker). It is then possible to diagnose the presence or absence of this mutation in II-3. If the result is negative, II-3 is then N/N and the risk for the child becomes null; there is also no need to analyse the mother. If the mutation is present, the risk $R1$ becomes equal to one, and the global risk for the child at birth climbs to 1 per cent (25 times more than the general population risk). It then becomes critical to determine the carrier status of II-4, possibly by systematic analyses using SSCP or DGGE (see Chapter 1 and Section 2.2).

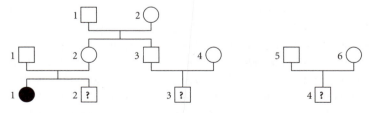

Figure 7.1

However, all of the PND methods in use today, even if they can produce individual solutions to identified couples at risk to their relatives, do not allow the resolution of the epidemiologic and public health problem which results from the fact that the majority of affected children are born to couples without a family history of the illness, like II-5 and II-6. To avoid such births and reduce the frequency of the disease significantly, it would be necessary to identify at-risk couples to be able to perform a PND for the first pregnancy, but a systematic identification of healthy carriers and at-risk couples poses technical, ethical and economic problems.

In fact, detection of an at-risk couple by molecular DNA analysis requires testing initially just a single member of the couple, and only testing the partner if the first is a healthy carrier. It would be useless to test the partner of an individual having a negative test result, whether this result was a true or false negative.

'False negative' individuals are non-identified healthy carriers, where the mutation they carry is either still unknown, or too rare to be tested for systematically (more than 100 different mutations are currently known for the CFTR gene). Define a detection rate H taking into account available knowledge and techniques to test the maximum number of mutations for a minimum time and cost. The complement $(1 - H)$ represents the failure rate for detection, that is to say the false negative rate, among a tested population of healthy carriers. A screen based only on testing the mutation $\Delta F508$, the most frequent, only detects 70 per cent of carriers ($H = 0.7$), whereas today it is possible to detect 80 per cent of carriers with a somewhat higher but still affordable cost.

The calculations carried out in the strategy diagram (Figure 7.2) show the results from using two detection rates $H = 0.70$ and $H = 0.90$. The failure of such a strategy is obvious since, of the four affected children in a cohort of 1000 couples ($1/2500 = 4/10\,000$), only 1.96 can be detected (in light grey, sensitivity 49 per cent) if $H = 0.70$. with 2.04 affecteds still remaining undetected (in dark grey) because one member of the couple was a carrier for a mutation not tested or not detectable. With a rate $H = 0.90$, the sensitivity only increases to 81 per cent (four affecteds detected out of five) whereas the cost to reach $H = 0.9$ becomes prohibitive.

In addition, such a strategy leads many couples to a distressing situation where one of them is a carrier and the other is negative, without knowing if this is a real or a false negative, with the paradox that the situation worsens as the detection rate increases (moving from 272 to 347 couples as H changes from 0.70 to 0.90). This situation poses an ethical problem as the average risk for these couples is definitely higher ($1/324$ if $H = 0.70$ and $1/964$ if $H = 0.90$) than the risk prior to the test ($1/2500$) with few possibilities to reassure those whose risk is really null and to inform those whose risk is in fact equal to $1/4$. It has been calculated that it is necessary to wait until the detection rate is 0.96 until the risk for the 'uncertain couples' drops to the population level of $1/2500$ and becomes ethically acceptable (the test not having a negative effect on the prior information). However, that remains for the moment economically impossible.

It is possible, however, that systematic testing of at-risk couples will become possible fairly quickly, partly because of rapid technological development (DNA chips)

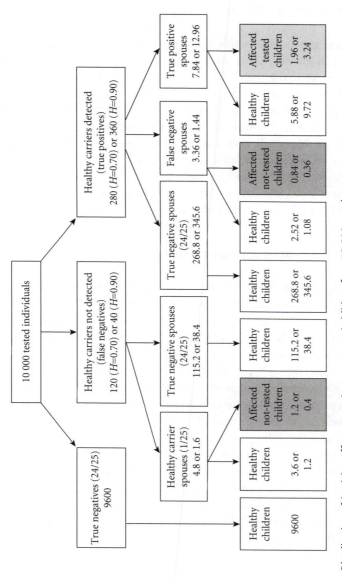

Figure 7.2   Distribution of healthy, affect, tested or non-tested children from 10 000 couples

reducing the cost of molecular analyses, and partly because of the possibility which seems to be developing to select the couples with the highest risk for whom a systematic test could be proposed (see Section 2.2).

7.5   Gametic disequilibrium

The study of the genetic composition of a population for a single gene is very restrictive, the more so as many traits are not monogenic but are dependent on multiple genes. However, studying the alleles of multiple genes or markers (cf. 'genetic fingerprints', Chapter 6), and of the diverse corresponding genotypes, the situation becomes quite complicated. A satisfactory example to illustrate this can be obtained by considering the genetic composition of a population for two diallelic autosomal genes or markers. The study of this situation allows the introduction of a new concept, gametic disequilibrium, to describe a reality which, if ignored, risks the drawing of false conclusions, equally well for basic research (gene mapping, analysis of the genetic diversity of populations, and the origins of certain mutations) as for applied research, notably in the calculation of risk associated with genetic fingerprints.

7.5.1   Allele frequencies and gamete frequencies

When studying genetic diversity relative to a single gene, the allele and gamete frequencies are always confounded, because even though alleles and gametes are not the same type of object they are, in this particular case, interchangeable: a gamete contains one allele of the gene, and the frequency of the gamete A1 is equal to the frequency of the allele A1.

As soon as a genetic study pertains to more than one gene, gametes and alleles are no longer interchangeable, and neither are allele and gamete frequencies. Consider the case of two diallelic genes A and B. For the first gene, define the alleles A1 with frequency p and A2 with frequency q. For the second gene, define the alleles B1 with frequency u and B2 with frequency v.

The gamete frequencies correspond to four different objects, with each one containing an allele from each of the two genes, that is to say the gametes (A1, B1) with frequency $g11$, (A1, B2) with frequency $g12$, (A2, B1) with frequency $g21$ and (A2, B2) with frequency $g22$, where the position in the index refers to the genes (gene A in the first position and B in the second), and the value of the index refers to the allelic type of the genes (value 1 for A1 or B1, value 2 for A2 or B2).

7.5.2   Gametic equilibrium and disequilibrium

A relationship clearly exists between the values of the gamete frequencies and the allele frequencies, p and q for gene A and u and v for gene B, but it is not obvious what this relationship is.

Imagine for the moment that the alleles in the two genes combine independently from each other in the gametes, with the selection of each allele being at random. In this case there is genetic equilibrium, and the four gamete frequencies are given by:

for   (A1, B1) $g11 = p \cdot u$
       (A1, B2) $g12 = p \cdot v$
       (A2, B1) $g21 = q \cdot u$
       (A2, B2) $g22 = q \cdot v$

However, this situation of gametic equilibrium is neither obligatory nor common in natural populations, even where there is Hardy–Weinberg equilibrium for the two genes separately (see below). When the situation of gametic equilibrium, for the two genes considered, is not present, the two genes are said to be in gametic disequilibrium. This is defined, for each of the four gametes, as the difference between the real frequency of a gamete, and the theoretical frequency for the gamete if it was in the state of gametic equilibrium, so:

for   (A1, B1) $\Delta 11 = g11 - p \cdot u$
       (A1, B2) $\Delta 12 = g12 - p \cdot v$
       (A2, B1) $\Delta 21 = g21 - q \cdot u$
       (A2, B2) $\Delta 22 = g22 - q \cdot v$

7.5.3 Origin of gametic disequilibrium

Numerous circumstances in the genetic history of the population can generate gametic disequilibrium, in fact all of the mechanisms which the model of Hardy–Weinberg assumes to be non-existent: mutations, migration, selection, genetic drift (which exists when the population size is small) and certain types of deviation from panmixia.

Generation of gametic disequilibrium following a migration

Two populations, even if they are close genetically, rarely have identical genetic compositions. In these conditions, migration between or mixture of the two populations will generate gametic disequilibrium between two polymorphic genes. This is a very common situation in humans when the populations are derived from a mixture of two or three continents (USA, Latin America, Mauritius, etc.).

Take the simple example of an equal mixture from two populations, A consisting exclusively of double homozygotes (A1/A1; B1/B1) and B consisting exclusively of double homozygotes (A2/A2; B2/B2). The resulting population will have allele frequencies equal to 1/4 for all of the alleles A1, A2, B1 and B2, but no individual will produce the gametes (A1, B2) or (A2, B1) – their frequencies will be zero, even though the product of their allele frequencies is equal to 1/4 for each of the two cases. Conversely, gametes (A1, B1) and (A2, B2) will have frequencies equal to 1/2, even though the product of their allele frequencies is only equal to 1/4. These differences between the gamete frequencies and the product of the frequencies of the alleles associated in the gametes illustrate gametic disequilibrium.

Clearly, in the next generation, the recombinant gametes (A1,B2) and (A2,B1) can be produced by A1B1/A2B2 individuals resulting in the first panmictic couples A1B1/A1B1 × A2B2/A2B2. This shows that gametic disequilibrium thus generated will progressively disappear under the effect of genetic recombination. However, it is clear that this disappearance will be slower the lower the frequency of genetic recombinations. In fact, if genes A and B are genetically independent (physically independent, or physically linked but with a large distance between them), A1B1/A2B2 individuals will make as many parental gametes (A1, B1) and (A2, B2) as recombinant gametes (A1, B2) and (A2, B1). In contrast, if genes A and B are strongly linked, then A1B1/A2B2 individuals will essentially only make the parental gametes (A1, B1) and (A2, B2), and very rarely the recombinant gametes (A1, B2) and (A2, B1). In this case, genetic disequilibrium is maintained for a long time and can be named, but only in this case, 'linkage disequilibrium'.

Generation of gametic disequilibrium following a mutation

Among the pathological mutations of the human gene CFTR implicated in cystic fibrosis (cf. Section 2.2), the most common, mutation ΔF508, is found in France in 70 per cent of healthy carriers. Upstream of the CFTR gene, exist two RFLP sites for the restriction enzymes *Taq* I and *Pst* I. According to whether the sites are present (+) or absent (−) at this position on chromosome 7, four associations in cis called 'haplotypes' can be defined (Table 7.4).

A chromosome 7 carrying the mutation ΔF508 is also carrying in 91.4 per cent of cases the haplotype B upstream of gene CFTR. In contrast, a chromosome 7 not carrying a mutation of CFTR only carries haplotype B in 19.4 per cent of cases. There exists, therefore, an association (non-functional but just physical or cartographic) between haplotype B on the one hand and the mutation ΔF508 on the other. This

Table 7.4 Frequency of each RFLP haplotype on the wild-type chromosomes in the French population and on the chromosomes carrying the mutation ΔF508

Taq I *Pst* I restriction sites		Haplotypes	Population haplotype frequency	Relative haplotype frequency on a wild-type chromosome	Relative haplotype frequency on a chromosome carrying ΔF508
−	−	A	32.6%	33.8%	1.3%
−	+	B	22.4%	19.4%	91.4%
+	−	C	33.6%	34.8%	0.5%
+	+	D	11.4%	12.1%	6.8%

association stems from the fact that the mutation ΔF508 (very old, its age is estimated at close to 40 000 years) occurred on a chromosome carrying haplotype B. The distance between the mutation and the RFLP sites is so small that crossovers between the two loci are very rare and the association between the mutation ΔF508 and the facultative sites on haplotype B has been maintained through time. This association constitutes a gametic disequilibrium also called, in this case, linkage disequilibrium.

The frequency, in the subset of gametes carrying the mutation ΔF508, of those who at the same time are carrying haplotype B, is equal to 91.4 per cent. The gametes carrying ΔF508 themselves represent 70 per cent of gametes carrying mutations of the CFTR gene, of which the global frequency in the general population represents 2 per cent (see above). The frequency of gametes carrying ΔF508 in the general population is therefore equal to

$$2\% \times 70\% = 1.4\%$$

The frequency of gametes carrying at the same time the mutation ΔF508 and haplotype B is then equal to

$$91.4\% \times 1.4\% = 1.27\%$$

At gametic equilibrium the frequency of these gametes would simply be equal to the product of 22.4 per cent (the general frequency of chromosomes carrying haplotype B) and 1.4 per cent (the frequency of chromosomes carrying ΔF508) giving 0.31 per cent. The gametic disequilibrium between the ΔF508 mutation and haplotype B is therefore equal to $1.27 - 0.31 = 0.96$ per cent.

7.5.4   Changes in gametic disequilibrium or linkage disequilibrium

As shown in the two previous examples, gametic disequilibrium tends to diminish under the effect of genetic recombination. This change is slower the tighter the linkage or the closer together the two genes are. It can be shown (see Serre, 1997) that disequilibrium changes following the equation:

$$\Delta_i = (1 - r)^i \Delta_0$$

where r is the recombination rate between the two genes (1/2 in the case of independence), i is the number of generations since the origin of the disequilibrium, and Δ_0 is the original value of the disequilibrium. It is therefore possible to estimate the time necessary to achieve equilibrium as a function of the recombination rate r between the two genes A and B.

It thus clearly appears, as the examples showed intuitively, that disequilibrium tends to zero over generations, and that this reduction depends strongly on the linkage or lack of genetic linkage between the two genes. If the genes are independent, r is equal to $1/2$ and $(1-r)^i$ tends very quickly to zero – in practice in eight to 10 generations. However, if the genes are tightly linked, the disequilibrium can persist for hundreds or thousands of generations. Gametic disequilibrium generated for two genes A and B, genetically independent (physically linked or not, this makes no difference), will have disappeared in eight to 10 generations (two centuries in humans). Gametic disequilibrium needs 60 generations if the genes are linked with a rate of 10 per cent; this represents 15 centuries in humans (the time since the baptism of Clovis!), and signifies that the disequilibrium generated between Gauls, Romans, Celts and other Germanic tribes should still persist for genes closer than 10 centi-Morgans (cM). As for disequilibrium existing between genes closer than 1 cM ($r < 0.01$), like the genes of the major histocompatability complex, this persists for thousands of years and could date back to the origin of modern man.

However, it is crucial when studying the genetic diversity of a population – particularly when using the method of genetic fingerprinting (cf. Chapter 6) – to have verified that the markers analysed are in gametic equilibrium. The calculation of the frequency of a genetic profile observed over several genes consists of the multiplication of the genotype frequency for each of them, which only corresponds to reality if the markers studied are in pairwise gametic equilibrium.

On this subject it is advisable to note that the fact that two genes are genetically independent, including those which are physically independent, is not a guarantee of gametic equilibrium: in the example of population mixture, which is a common situation in human populations, the genes A1–A2 and B1–B2 could be physically independent and the mixture would generate disequilibrium. What changes with genetic independence, is the speed with which equilibrium is reached, it is faster, but 10 generations is still two centuries!

7.6   Reference and Bibliography

7.6.1   Reference

Serre, J.-L. (1997). *Génétique des populations: modèles de bases et applications.* Nathan-Université, Paris.

7.6.2   Bibliography

Gouyon, P.-H. (1998). *Précis de génétique des populations.* Dunod, Paris.
Solignac, M. *et al.* (1995). *Génétique et evolution.* Hermann, Paris.

Index